Short Handbook of Mathematical Formulas for Chemical Engineers

Muhammad Rashid Usman

Department of Petroleum and Chemical Engineering
Sultan Qaboos University, Muscat
Institute of Chemical Engineering and Technology
University of the Punjab, Lahore

Copyright © 2017 Muhammad Rashid Usman
All rights reserved.
ISBN-13: 978-1976404467
ISBN-10: 1976404460

To my daughter Maryam

Preface

This book is a collection of physical constants, conversion factors, mathematical formulas, and some essential MATLAB commands many of which are required on daily basis by chemical engineers. The book is intended to save vital time of the students, instructors, researchers, and people in the process industry which they have to spend in searching the material available in this book.

The author hopes that the reader will find this book a useful reference and continue to use it for the solution to his or her problems.

The author acknowledges his family members for their patience and support.

Table of Contents

1.0	Greek Alphabetic Symbols	1
2.0	Physical Constants	2
2.1	Universal Molar Gas Constant, R, in Different Units	2
2.2	Standard Acceleration Due to Gravity, g	3
2.3	Additional Physical Constants	3
2.4	SI Unit Prefixes	3
3.0	Commonly Employed Units with Their Corresponding System of Units	4
4.0	Conversion of Units	5
4.1	Conversion Factor, g_c	5
4.2	Units of Angle	5
4.3	Units of Area	5
4.4	Units of Concentration	5
4.5	Units of Density or Mass Concentration	6
4.5.1	Specific Gravity Scales	6
4.5.1.1	Specific Gravity, Common	6
4.5.1.2	Baumé Gravity Scale	6
4.5.1.3	API Gravity Scale	7
4.6	Units of Diffusivity (Momentum, Mass, or Thermal Diffusivity)	7
4.7	Units of Energy, Heat, or Work	7
4.8	Units of Force	7
4.9	Units of Heat Capacity, Specific	8
4.10	Units of Heat Flux	8
4.11	Units of Heat Transfer Coefficient	8
4.12	Units of Kinematic Viscosity or Momentum Diffusivity	8
4.12.1	Different Scales for Kinematic Viscosity Used with Industrial Viscometers	8
4.13	Units of Length	9
4.14	Units of Mass	10
4.15	Units of Mass Flowrate or Rate of Mass Transfer	10
4.16	Units of Mass Transfer Coefficient	10
4.16.1	With Concentration Difference	10
4.16.2	With Mole Fraction Difference	10
4.16.3	With Partial Pressure Difference	11
4.17	Units of Mass Velocity or Mass Flux	11
4.18	Units of Molar Concentration	11
4.19	Units of Molar Flowrate or Molar Rate of Mass Transfer	11
4.20	Units of Molar Flux	12
4.21	Units of Momentum	12
4.22	Units of Momentum Flux or Shear Stress	12

4.23	Units of Power or Rate of Energy Transfer or Rate of Heat Transfer	12
4.24	Units of Pressure or Momentum Flux or Shear Stress	13
4.25	Units of Rate of Chemical Reaction	13
4.25.1	In Terms of Volume of Reactor	13
4.25.2	In Terms of Weight of Reactor	13
4.26	Units of Space Velocity	14
4.27	Units of Temperature	14
4.28	Units of Thermal Conductivity	14
4.29	Units of Time	15
4.30	Units of Velocity, Linear	15
4.31	Units of Viscosity (Absolute or Dynamic)	15
4.32	Units of Volume	16
4.33	Units of Volumetric Flowrate	16
5.0	Indices	17
6.0	Logarithms	18
7.0	Facts from Algebra	19
7.1	Binomial Theorem	19
7.2	Permutations and Combinations	19
7.2.1	Permutations	19
7.2.2	Combinations	20
7.3	Quadratic Equation	20
7.4	Cubic Equation	21
8.0	Trigonometry	23
8.1	Solution of a Triangle	24
8.1.1	Law of Sines	25
8.1.2	Law of Cosines	25
8.1.3	Law of Tangents	25
8.1.4	Half Angle Formulas	25
8.2	Solution of a Right-Angled Triangle	26
8.3	Area of a Triangle	26
8.4	Circumradius of a Triangle	27
8.5	Inradius of a Triangle	27
8.6	Inverse Trigonometric Functions	28
8.7	Hyperbolic Functions	28
8.8	Inverse Hyperbolic Functions	29
9.0	Series	30
9.1	Arithmetic Series	30
9.2	Geometric Series	30
9.3	Binomial Series	31
9.4	Taylor Series	31

10.0	Matrices	32
10.1	Addition of Matrices A and B	32
10.2	Subtraction of Matrices: Matrix B from Matrix A	33
10.3	Multiplication of Matrix A by a Scalar	33
10.4	Multiplication of Matrix A and Matrix B	33
10.5	Determinant of Matrix A	34
10.6	Inverse of Matrix A	34
10.7	Transpose of Matrix A	34
10.8	Solution of Simultaneous Equations by Matrix Inverse Method	35
10.9	Cramer Rule	36
10.10	Eigen Values and Eigen Vectors	37
11.0	Complex Numbers	38
12.0	Plane Geometry	40
12.1	Square	40
12.2	Rectangle	40
12.3	Rhombus	40
12.4	Parallelogram	40
12.5	Trapezoid	40
12.6	Triangle	41
12.6.1	Right-Angled Triangle	41
12.6.2	Equilateral Triangle	41
12.7	Polygon	41
12.7.1	Polygon, Irregular	41
12.7.1.1	Quadrilateral, Irregular	42
12.7.2	Polygon, Regular	42
12.8	Circle	42
12.9	Annulus	43
12.10	Sector of a Circle	43
12.11	Sector of an Annulus	43
12.12	Segment of a Circle	44
12.13	Ellipse	44
12.14	Segment of a Parabola	44
13.0	Solid Geometry	45
13.1	Cube	45
13.2	Rectangular Parallelepiped or Cuboid	45
13.3	Parallelepiped	45
13.4	Cylinder, Solid	45
13.5	Cylinder, Hollow	46
13.6	Sliced Cylinder or Cylinder with Oblique Face	46
13.7	Ungula or Cylindrical Hoof	46

13.8	Cylinder with Slant Height	47
13.9	Cone	47
13.10	Frustum of a Cone	47
13.11	Prism	47
13.11.1	Regular Triangular Prism	48
13.11.2	Square Prism	48
13.11.3	Prism with Polygon as Base	48
13.12	Pyramid	48
13.12.1	Triangular Pyramid	48
13.12.2	Square Pyramid	49
13.12.3	Pyramid with Polygon as Base	49
13.13	Frustum of a Pyramid or Cone	49
13.14	Barrel	49
13.15	Sphere	49
13.16	Zone of a Sphere	50
13.17	Segment of a Sphere or Spherical Cap	50
13.18	Sector of a Sphere	50
13.19	Sphere with Cylindrical Boring	50
13.20	Spherical Triangle	50
13.21	Torus or Anchor Ring	51
13.22	Ellipsoid	51
13.23	Paraboloid	51
14.0	Analytical Geometry	52
14.1	Distance between Two Points on a Straight Line	52
14.2	Mid-Point of Two Points on a Straight Line	52
14.3	Slope of Straight Line	52
14.4	Equation of a Straight Line	52
14.4.1	Two-Point Form	52
14.4.2	Point-Slope Form	52
14.4.3	Slope-Intercept Form	52
14.4.4	Intercepts Form	53
14.4.5	Normal Form	53
14.4.6	Line Parallel to the x-axis	53
14.4.7	Line Parallel to the y-axis	53
14.4.8	General Equation of a Straight Line	53
14.5	Ratio Formula	53
14.6	Condition of Parallelism and Perpendicularity of Two Lines	54
14.7	Condition of Concurrency of Three Lines	55
14.8	Distance of a Point from a Line	55
14.9	Area of a Triangular Region	56

14.10	Condition of Collinearity (on a Single Line) of Three Points	56
14.11	Circle	56
14.12	Ellipse	57
14.13	Parabola	58
14.14	Hyperbola	59
15.0	Derivatives	60
16.0	Integrals	63
16.1	Area under the Curve	65
17.0	Vectors	66
17.1	Dot Product of Vectors	67
17.2	Cross Product of Vectors	67
17.3	Scalar Triple Product of Vectors	68
17.4	Vector Triple Product of Vectors	68
18.0	Coordinate Systems	69
18.1	Rectangular or Cartesian Coordinates	69
18.2	Cylindrical Coordinates	69
18.3	Spherical Coordinates	69
19.0	Gradient, Divergence, Curl, and Laplacian	70
19.1	Gradient, Divergence, Curl, and Laplacian in Different Coordinate Systems	70
20.0	Laplace Transform	71
20.1	Table of Laplace Transforms	71
21.0	Statistics	72
21.1	Averages	72
21.1.1	Arithmetic Mean or Mean	72
21.1.1.1	Weighted Arithmetic Mean	72
21.1.2	Geometric Mean	72
21.1.3	Harmonic Mean	73
21.1.4	Relationship between Arithmetic, Geometric, and Harmonic Means	73
21.1.5	Root Mean Square or Quadratic Mean	73
21.1.6	Logarithmic Mean or Log Mean	74
21.2	Median	74
21.3	Mode	75
21.4	Measures of Dispersion	75
21.4.1	Range	75
21.4.2	Mean or Average Deviation	75
21.4.3	Standard Deviation	76
21.4.4	Variance	76
21.4.5	Standard Error of Mean	77
21.5	Measures of Error of Estimation	78
21.5.1	Sum of Squares of the Errors (*SSE*)	78

21.5.2	R^2	78
21.5.3	AdjR^2	78
21.5.4	F-value	78
21.5.5	Root Mean Square Deviation	79
21.5.6	Standard Error of Estimate	79
21.6	Curve Fitting Models	79
21.7	Statistical Distributions	80
22.0	Mathematical Functions	81
22.1	Gamma, Beta, Error, and Complementary Error Functions	81
22.2	Values of Error Function	81
23.0	Numerical Mathematics	82
23.1	Interpolation	82
23.1.1	Linear	82
23.1.2	Quadratic	82
23.1.3	Lagrange	82
23.1.3.1	First Order Lagrange Polynomial	82
23.1.3.2	Second Order Lagrange Polynomial	83
23.2	Solution of Algebraic Equations	83
23.2.1	Newton-Raphson Method	83
23.2.2	Secant Method	83
23.3	Solution of First Order Ordinary Differential Equations (ODEs)	83
23.3.1	Euler Method	83
23.3.2	Improved Euler Method: Heun Method	83
23.3.3	Runge-Kutta (RK) Method, 4th Order	84
23.3.4	Runge-Kutta-Fehlberg (RKF) Method	84
23.4	Differentiating a Function: Finite Differences	85
23.4.1	Forward Finite Differences	85
23.4.2	Backward Finite Differences	85
23.4.3	Centered Finite Differences	86
23.5	Integrating a Function	86
23.5.1	Trapezoidal Rule	86
23.5.2	Simpson 1/3 Rule	86
23.6	Partial Differential Equations	86
23.6.1	Classification of Second Order Linear Partial Differential Equations	86
23.6.2	Explicit Finite Difference Method for One-Dimensional Parabolic Differential Equation	87
24.0	Selected MATLAB Commands with Examples	89
25.0	Suggested Sources	94

1.0 Greek Alphabetic Symbols

Upper Case	Lower Case	Name
A	α	Alpha
B	β	Beta
Γ	γ	Gamma
Δ	δ	Delta
E	$\varepsilon\ \epsilon$	Epsilon
Z	ζ	Zeta
H	η	Eta
Θ	$\theta\ \vartheta$	Theta
I	ι	Iota
K	κ	Kappa
Λ	λ	Lambda
M	μ	Mu
N	ν	Nu
Ξ	ξ	Xi
O	o	Omicron
Π	π	Pi
P	$\rho\ \varrho$	Rho
Σ	$\sigma\ \varsigma$	Sigma
T	τ	Tau
Y ϒ	υ	Upsilon
Φ	$\varphi\ \phi$	Phi
X	χ	Chi
Ψ	ψ	Psi
Ω	ω	Omega

2.0 Physical Constants

2.1 Universal Molar Gas Constant, *R*, in Different Units

Unit				Value
Volume	Pressure	Mole	Temperature	
m^3	Pa	mol	K	8.3145
m^3	Pa	kmol	K	8314.5
m^3	kPa	mol	K	0.0083145
m^3	kPa	kmol	K	8.3145
m^3	bar	mol	K	8.314×10^{-5}
m^3	bar	kmol	K	0.083145
m^3	atm	kmol	K	0.08206
m^3	kg_f/cm^2	kmol	K	0.08478
cm^3	kPa	mol	K	8314.5
cm^3	bar	mol	K	83.145
cm^3	atm	gmol	K	82.057
L	atm	mol	K	0.08206
ft^3	psia	lbmol	°R	10.732
ft^3	atm	lbmol	°R	0.73024
Energy				
J		mol	K	8.3145
J		kmol	K	8314.5
kJ		mol	K	0.0083145
kJ		kmol	K	8.3145
cal(IT)		gmol	K	1.9859
cal(TC)		gmol	K	1.9872
Btu(IT)		lbmol	°R	1.9859
$ft \cdot lb_f$		lbmol	°R	1545.3

IT means international steam table, 1 Btu(IT) = 1055.056 J, 1 cal(IT) = 4.1868 J and TC means thermochemical, 1 cal(TC) = 4.184 J.

Ideal gas law (Eq. 2.1) can be used to calculate the value of *R* in a required set of units:

2.1) $R = \dfrac{pV}{nT}$

where R = gas constant, p = absolute pressure, V = total volume occupied, n = total number of moles occupying volume V, and T = absolute temperature.

2.2 Standard Acceleration Due to Gravity, g

Unit	Numerical Value	Commonly Used Value
m/s^2	9.8067	9.81
cm/s^2	980.665	981
ft/s^2	32.174	32.2

2.3 Additional Physical Constants

Name	Symbol	Value	SI unit
Atomic mass unit	m_u	1.661×10^{-27}	kg
Avogadro number	N_A	6.022×10^{23}	mol^{-1}
Boltzmann constant	$k_B = R/N_A$	1.381×10^{-23}	J/K
Gravitational constant	G	6.674×10^{-11}	N·m^2/kg^2
Permeability of free space	μ_o	$4\pi \times 10^{-7}$	H/m
Permittivity of free space	ϵ_o	8.854×10^{-12}	F/m
Planck constant	h	6.626×10^{-34}	J·s
Speed of light in vacuum	c	2.998×10^8	m/s
Stefan-Boltzmann constant	σ	5.670×10^{-8}	W/(m^2·K^4)
Velocity of sound in dry air at 0°C and 1 atm	v	331.5	m/s
Wien displacement constant	b	2.898×10^{-3}	m·K

2.4 SI Unit Prefixes

Prefix	Multiplier	Symbol	Prefix	Multiplier	Symbol
deca	10^1	da	deci	10^{-1}	d
hecto	10^2	h	centi	10^{-2}	c
kilo	10^3	k	milli	10^{-3}	m
mega	10^6	M	micro	10^{-6}	μ
giga	10^9	G	nano	10^{-9}	n
tera	10^{12}	T	pico	10^{-12}	p
peta	10^{15}	P	femto	10^{-15}	f
exa	10^{18}	E	atto	10^{-18}	a
zetta	10^{21}	Z	zepto	10^{-21}	z
yotta	10^{24}	Y	yocto	10^{-24}	y

3.0 Commonly Employed Units with their Corresponding System of Units

Quantity	American Engineering System (AE)		cgs System		International System of Units (SI)	
	Unit	Symbol	Unit	Symbol	Unit	Symbol
Length	foot	ft	centimeter	cm	meter	m
Mass	pound mass	lb_m	gram	g	kilogram	kg
Amount of substance	pound mole	lbmol	gram mole	gmol	mole	mol
Time	second hour	s h	second	s	second	s
Temperature	Fahrenheit Rankine	°F °R	Celsius Kelvin	°C K	Celsius Kelvin	°C K
Force	pound force	lb_f	dyne	dyn	Newton	N
Energy	foot-pound force British thermal unit	$ft·lb_f$ Btu	dyne-centimeter erg calorie	dyn·cm erg cal	Newton-meter Joule	N·m J
Power	foot-pound force/second British thermal unit per hour Horse power	$ft·lb_f/s$ Btu/h hp	dyne-centimeter/ second erg per second calorie per second	dyn·cm/s erg/s cal/s	Newton-meter/ second Watt	N·m/s W
Pressure	pound force/ square inch	lb_f/in^2	dyne/square centimeter	dyn/cm^2	Newton/square meter Pascal	N/m^2 Pa
Density	pound mass/ cubic foot	lb_m/ft^3	gram/cubic centimeter	g/cm^3	kilogram/cubic meter	kg/m^3
Velocity	foot/second	ft/s	centimeter/ second	cm/s	meter/second	m/s

o In American Engineering System both mass and force are basic quantities and 1 pound force is a force that produces an acceleration of 32.174 ft/s² in 1 pound mass body. Therefore, the conversion factor g_c is required as its value is not unity. See Section 4.1.

o In British Engineering System force is a basic quantity but not the mass. Here, 1 pound force is a force that produces an acceleration of 1 ft/s² in mass of 1 slug body. Therefore, there is no need to use the conversion factor g_c as its value is unity. See Section 4.1.

o English Absolute System or fps (foot-pound-second) System uses poundal as unit of force where 1 poundal is a force that produces an acceleration of 1 ft/s² in 1 pound mass body. Again, there is no need to use the conversion factor g_c as its value is unity.

4.0 Conversion of Units

4.1 Conversion Factor, g_c

System of Units	Units	Numerical Value
American Engineering	$lb_m \cdot ft/(lb_f \cdot s^2)$	32.174
British Engineering	$lb_m \cdot ft/(lb_f \cdot s^2)$	1.0
cgs	$g \cdot cm/(dyn \cdot s^2)$	1.0
SI	$kg \cdot m/(N \cdot s^2)$	1.0

lb_m is lb, avoir

4.2 Units of Angle

deg	min	rad	rev	s
1	60	0.017453	0.0027778	3600
0.016667	1	2.909×10^{-4}	4.630×10^{-5}	60
57.296	3437.7	1	0.15915	206264.8
360	21600	6.2832	1	1296000
2.778×10^{-4}	0.016667	4.848×10^{-6}	7.716×10^{-7}	1

1 grad = 0.9 deg   180 deg = π rad

4.3 Units of Area

in^2	ft^2	$yard^2$	cm^2	m^2
1	0.0069444	7.716×10^{-4}	6.4516	6.452×10^{-4}
144	1	0.11111	929.03	0.092903
1296	9	1	8361.3	0.83613
0.15500	0.0010764	1.196×10^{-4}	1	0.0001
1550.0	10.764	1.1960	10000	1

1 acre = 4840 $yard^2$ = 4046.9 m^2 1 hectare = 10000 m^2
1 km^2 = 10^6 m^2 1 $mile^2$ = 2.5900 km^2 = 2.590×10^6 m^2

4.4 Units of Concentration

See units of molar concentration (Section 4.18).

For mass concentration see units of density (Section 4.5).

4.5 Units of Density or Mass Concentration

lb_m/ft^3	$lb_m/gal(US)$	g/cm^3	kg/m^3	g/L
1	0.13368	0.016018	16.018	16.018
7.4805	1	0.11983	119.83	119.83
62.428	8.3454	1	1000	1000
0.062428	0.0083454	0.001	1	1

$1\ lb_m/in^3 = 1728\ lb_m/ft^3$
$1\ kg/L = 1\ g/cm^3$

$1\ g/dm^3 = 1\ g/L$
$1\ slug/ft^3 = 32.174\ lb_m/ft^3$

4.5.1 Specific Gravity Scales

4.5.1.1 Specific Gravity, Common

4.1) $s = \dfrac{\text{density of material under study}}{\text{density of reference material}}$

Water is usually the reference material in the case of liquids and solids, while air is the reference gas in the case of gases. For oil industry it is usually $s\dfrac{60°F}{60°F}$ showing density of material and reference both at 60°F. For scientific purposes it is usually $s\dfrac{T°C}{4°C}$ where T is the temperature at which density measurement of a material is carried out.

4.5.1.2 Baumé Gravity Scale

For liquids less dense than water:

4.2) $°Be' = \dfrac{140}{s^{60°F}_{60°F}} - 130$

4.3) $s^{60°F}_{60°F} = \dfrac{140}{°Be' + 130}$

For liquids more dense than water:

4.4) $°Be' = 145 - \dfrac{145}{s^{60°F}_{60°F}}$

4.5) $s^{60°F}_{60°F} = \dfrac{145}{145 - °Be'}$

4.5.1.3 API Gravity Scale

4.6) $°API = \dfrac{141.5}{s_{60°F}^{60°F}} - 131.5$ 4.7) $s_{60°F}^{60°F} = \dfrac{141.5}{°API + 131.5}$

4.6 Units of Diffusivity (Momentum, Mass, or Thermal Diffusivity)

ft²/h	ft²/s	cm²/s = St	m²/s	1 mm²/s = cSt
1	2.778×10^{-4}	0.25806	2.581×10^{-5}	25.806
3600	1	929.03	0.092903	92903.0
3.8750	0.0010764	1	0.0001	100
38750.1	10.764	10000	1	1×10^{6}
0.038750	1.076×10^{-5}	0.01	1×10^{-6}	1

1 m²/h = 277.78 mm²/s   1 in²/s = 25 ft²/h

4.7 Units of Energy, Heat, or Work

ft·lb_f	Btu(IT)	cal(TC)	J (or N·m)	kJ
1	0.0012851	0.32405	1.3558	0.0013558
778.17	1	252.16	1055.1	1.0551
3.0860	0.0039657	1	4.184	0.004184
0.73756	9.478×10^{-4}	0.23901	1	0.001
737.56	0.94782	239.01	1000	1

1 cal(IT) = 4.1868 J   1 Btu(TC) = 1054.35 J
1 erg = 1×10^{-7} J   1 therm(US) = 1.055×10^{8} J
1 kWh = 3.6×10^{6} J   1 MMBtu = 1×10^{6} Btu
1 toe (tonne of oil equivalent) = 4.187×10^{10} J

4.8 Units of Force

lb_f	N	kg_f	poundal	dyn
1	4.4482	0.45359	32.174	444822.2
0.22481	1	0.10197	7.2330	100000
2.2046	9.8067	1	70.932	980665
0.031081	0.13825	0.014098	1	13825.5
2.248×10^{-6}	1×10^{-5}	1.020×10^{-6}	7.233×10^{-5}	1

4.9 Units of Heat Capacity, Specific*

Btu(IT)/(lb$_m$·°F)	J/(g·°C)	kJ/(kg·°C)	cal(TC)/(g·°C)	cal(IT)/(g·°C)
1	4.1868	4.1868	1.0007	1
0.23885	1	1	0.23901	0.23885
0.99933	4.184	4.184	1	0.99933

*°C can be replaced with K and °F can be replaced with °R without affecting the value.

4.10 Units of Heat Flux

Btu(IT)/(h·ft^2)	cal(TC)/(s·cm^2)	W/cm^2	W/m^2	kW/m^2
1	7.540×10^{-5}	3.155×10^{-4}	3.1546	0.0031546
13263.2	1	4.184	41840	41.84
3170.0	0.23901	1	10000	10
0.31700	2.390×10^{-5}	0.0001	1	0.001
317.00	0.023901	0.1	1000	1

1 Btu(IT)/(s·ft^2) = 11356.5 W/m^2

4.11 Units of Heat Transfer Coefficient*

Btu(IT)/(h·ft^2·°F)	W/(cm^2·°C)	W/(m^2·°C)	kW/(m^2·°C)	cal(TC)/(s·cm^2·°C)
1	5.678×10^{-4}	5.6783	0.0056783	1.357×10^{-4}
1761.1	1	10000	10	0.23901
0.17611	0.0001	1	0.001	2.390×10^{-5}
176.11	0.10	1000	1	0.023901
7368.5	4.184	41840	41.840	1

*°C can be replaced with K and °F can be replaced with °R without affecting the value.

4.12 Units of Kinematic Viscosity or Momentum Diffusivity

Units are those of diffusivity. See Section 4.6.

4.12.1 Different Scales for Kinematic Viscosity Used with Industrial Viscometers

The general formula is

$$4.8) \text{ Kinematic viscosity} = \frac{\text{absolute viscosity}}{\text{density}} = v = At - \frac{B}{t}$$

The values of A and B should be found by calibrating the corresponding equipment. However, the following guidelines are useful.

Eq.	Scale	Kinematic Viscosity (St)	Efflux Time, t (s)
4.9	Engler	$0.00147t - \dfrac{3.74}{t}$	
4.10	Redwood Admiralty	$0.027t - \dfrac{20}{t}$	
4.11	Redwood No. 1	$0.00260t - \dfrac{1.79}{t}$	$34 < t < 100$
4.12	Redwood No. 1	$0.00247t - \dfrac{0.50}{t}$	$t > 100$
4.13	Saybolt Furol	$0.0224t - \dfrac{1.84}{t}$	$25 < t < 40$
4.14	Saybolt Furol	$0.0216t - \dfrac{0.60}{t}$	$t > 40$
4.15	Saybolt Universal	$0.00226t - \dfrac{1.95}{t}$	$32 < t < 100$
4.16	Saybolt Universal	$0.00220t - \dfrac{1.35}{t}$	$t > 100$

Ref.: Green, D.W.; Perry, R.H. 2008. Perry's Chemical Engineers' Handbook. 8th ed. McGraw-Hill, New York, p. 1-17.

4.13 Units of Length

in	ft	yard	mm	cm	m	km
1	0.083333	0.027778	25.4	2.54	0.0254	2.54×10^{-5}
12	1	0.33333	304.8	30.48	0.3048	3.048×10^{-4}
36	3	1	914.4	91.44	0.9144	9.144×10^{-4}
0.039370	0.0032808	0.0010936	1	0.1	0.001	1×10^{-6}
0.39370	0.032808	0.010936	10	1	0.01	1×10^{-5}
39.370	3.2808	1.0936	1000	100	1	0.001
39370.1	3280.8	1093.6	1×10^6	100000	1000	1

1 micron = 1×10^{-6} m
1 Å = 1×10^{-8} cm = 1×10^{-10} m
1 mile = 1609.3 m = 1.6093 km
1 fathom = 6 ft

1 μm = 1×10^{-6} m
1 nm = 1×10^{-9} m
1 mil = 0.001 in
1 furlong = 660 ft

4.14 Units of Mass

lb_m	mg	g	kg	tonne (metric ton)
1	453592.4	453.59	0.45359	4.536×10^{-4}
2.205×10^{-6}	1	0.001	1×10^{-6}	1×10^{-9}
0.0022046	1000	1	0.001	1×10^{-6}
2.2046	1×10^6	1000	1	0.001
2204.6	1×10^9	1×10^6	1000	1

1 carat, metric = 200 mg
1 slug = 14.594 kg
1 grain = 0.064799 g
1 oz, troy = 1.0971 oz, avoir = 31.103 g

1 ton, short = 2000 lb_m
1 ton, long = 2240 lb_m
1 oz, avoir = 28.350 g
1 lb_m, troy = 0.82286 lb_m, avoir

4.15 Units of Mass Flowrate or Rate of Mass Transfer

lb_m/s	lb_m/h	g/s	kg/s	kg/h	tonne (metric ton)/day
1	3600	453.59	0.45359	1632.9	39.190
2.778×10^{-4}	1	0.12600	1.260×10^{-4}	0.45359	0.010886
0.0022046	7.9366	1	0.001	3.6	0.0864
2.2046	7936.6	1000	1	3600	86.4
6.124×10^{-4}	2.2046	0.27778	2.778×10^{-4}	1	0.024
0.025516	91.859	11.574	0.011574	41.667	1

1 lb_m/min = 0.0075599 kg/s   1 g/h = 2.778×10^{-7} kg/s
1 mg/h = 2.778×10^{-10} kg/s
1 tonne (metric ton)/year = 3.171×10^{-5} kg/s (based on 1 year = 365 day).
The unit tonne/year is commonly based on stream days, i.e., working days of a plant in a year.

4.16 Units of Mass Transfer Coefficient

4.16.1 With Concentration Difference

Units are those of velocity. See Section 4.30.

4.16.2 With Mole Fraction Difference

Units are those of molar flux. See Section 4.20.

4.16.3 With Partial Pressure Difference

lbmol/(h·ft²·psi)	gmol/(s·cm²·atm)	mol/(s·m²·bar)	mol/(s·m²·Pa)	mol/(s·m²·kPa)
1	0.0019931	19.670	1.967×10^{-4}	0.19670
501.73	1	9869.2	0.098692	98.692
0.050838	1.013×10^{-4}	1	1×10^{-5}	0.01
5083.8	10.133	100000	1	1000
5.0838	0.010133	100	0.001	1

1 kmol/(s·m²·kPa) = 1 mol/(s·m²·Pa)

4.17 Units of Mass Velocity or Mass Flux

lb_m/(s·ft²)	lb_m/(h·ft²)	g/(s·cm²)	g/(s·m²)	kg/(s·m²)	kg/(h·m²)
1	3600	0.48824	4882.4	4.8824	17576.7
2.778×10^{-4}	1	1.356×10^{-4}	1.3562	0.0013562	4.8824
2.0482	7373.4	1	10000	10	36000
2.048×10^{-4}	0.73734	0.0001	1	0.001	3.6
0.20482	737.34	0.1	1000	1	3600
5.689×10^{-5}	0.20482	2.778×10^{-5}	0.27778	2.778×10^{-4}	1

4.18 Units of Molar Concentration

lbmol/ft³	gmol/cm³	mol/m³	kmol/m³	mol/L = mol/dm³
1	0.016018	16018.5	16.018	16.018
62.428	1	1×10^6	1000	1000
6.243×10^{-5}	1×10^{-6}	1	0.001	0.001
0.062428	0.001	1000	1	1

1 lbmol/in³ = 1728 lbmol/ft³   1 mmol/cm³ = 1000 mol/m³ = 1 mol/L

4.19 Units of Molar Flowrate or Molar Rate of Mass Transfer

lbmol/s	lbmol/h	mol/s	kmol/s	kmol/h
1	3600	453.59	0.45359	1632.9
2.778×10^{-4}	1	0.12600	1.260×10^{-4}	0.45359
0.0022046	7.9366	1	0.001	3.6
2.2046	7936.6	1000	1	3600
6.124×10^{-4}	2.2046	0.27778	2.778×10^{-4}	1

4.20 Units of Molar Flux

lbmol/(s·ft^2)	lbmol/(h·ft^2)	gmol/(s·cm^2)	mol/(s·m^2)	kmol/(s·m^2)	kmol/(h·m^2)
1	3600	0.48824	4882.4	4.8824	17576.7
2.778×10^{-4}	1	1.356×10^{-4}	1.3562	0.0013562	4.8824
2.0482	7373.4	1	10000	10	36000
2.048×10^{-4}	0.73734	0.0001	1	0.001	3.6
0.20482	737.34	0.1	1000	1	3600
5.689×10^{-5}	0.20482	2.778×10^{-5}	0.27778	2.778×10^{-4}	1

4.21 Units of Momentum

lb$_m$·ft/s	lb$_m$·ft/h	g·cm/s	kg·m/s
1	3600	13825.5	0.13825
2.778×10^{-4}	1	3.8404	3.840×10^{-5}
7.233×10^{-5}	0.26039	1	1×10^{-5}
7.2330	26038.9	100000	1

4.22 Units of Momentum Flux or Shear Stress

Units are those of pressure. See Section 4.24.

4.23 Units of Power or Rate of Energy or Rate of Heat Transfer

ft·lb$_f$/s	ft·lb$_f$/min	J/s = W	kJ/s = kW	kJ/h	Btu(IT)/h	hp (British)
1	60	1.3558	0.0013558	4.8809	4.6262	0.0018182
0.016667	1	0.022597	2.260×10^{-5}	0.081349	0.077104	3.030×10^{-5}
0.73756	44.254	1	0.001	3.6	3.4121	0.0013410
737.56	44253.7	1000	1	3600	3412.1	1.3410
0.20488	12.293	0.27778	2.778×10^{-4}	1	0.94782	3.725×10^{-4}
0.21616	12.969	0.29307	2.931×10^{-4}	1.0551	1	3.930×10^{-4}
550	33000	745.70	0.74570	2684.5	2544.4	1

1 Btu(IT)/s = 778.17 ft·lb$_f$/s = 1055.1 W   1 Btu(IT)/min = 17.584 W
1 cal(TC)/s = 4.184 W   1 cal(IT)/s = 4.1868 W
1 kcal(TC)/h = 1.1622 W   1 erg/s = 1×10^{-7} W
1 MW = 1000 kW = 1×10^6 W   1 GW = 1×10^9 W
1 ton (refrigeration) = 12000 Btu(IT)/h = 3516.9 W
1 MMBtu/h = 1×10^6 Btu/h

4.24 Units of Pressure or Momentum Flux or Shear Stress

psi (lb$_f$/in^2)	kg$_f$/cm^2	Pa = N/m^2	kPa	bar	atm	mmHg (0°C)
1	0.070307	6894.8	6.8948	0.068948	0.068046	51.715
14.223	1	98066.5	98.067	0.98067	0.96784	735.56
1.450×10^{-4}	1.020×10^{-5}	1	0.001	1×10^{-5}	9.869×10^{-6}	0.0075006
0.14504	0.010197	1000	1	0.01	0.0098692	7.5006
14.504	1.0197	100000	100	1	0.98692	750.06
14.696	1.0332	101325	101.325	1.01325	1	760.00
0.019337	0.0013595	133.32	0.13332	0.0013332	0.0013158	1

1 MPa = 1000 kPa = 1×10^6 Pa = 10 bar   1 dyn/cm^2 = 0.1 Pa
1 lb$_f$/ft^2 = 47.880 Pa   1 torr = 133.32 Pa   (1 atm = 760.00 torr)
1 mbar = 100 Pa
1 inH$_2$O (4°C) = 249.08 Pa   (1 atm = 406.79 inH$_2$O at 4°C)
1 ftH$_2$O (4°C) = 2989.0 Pa   (1 atm = 33.900 ftH$_2$O at 4°C)
1 inHg (0°C) = 3386.4 Pa   (1 atm = 29.921 inHg at 0°C)

4.17) Absolute pressure = atmospheric pressure + gauge pressure
4.18) Absolute pressure = atmospheric pressure – vacuum

4.25 Units of Rate of Chemical Reaction

4.25.1 In Terms of Volume of Reactor

lbmol/(ft^3·s)	lbmol/(ft^3·h)	gmol/(cm^3·s)	mol/(m^3·s)	kmol/(m^3·h)
1	3600	0.016018	16018.5	57666.5
2.778×10^{-4}	1	4.450×10^{-6}	4.4496	16.018
62.428	224740.7	1	1×10^6	3.6×10^6
6.243×10^{-5}	0.22474	1×10^{-6}	1	3.6
1.734×10^{-5}	0.062428	2.778×10^{-7}	0.27778	1

4.25.2 In Terms of Weight of Catalyst

lbmol/(lb$_m$·s)	lbmol/(lb$_m$·h)	gmol/(g·s)	mol/(kg·s)	kmol/(kg·h)
1	3600	1	1000	3600
2.778×10^{-4}	1	2.778×10^{-4}	0.27778	1
0.001	3.6	0.001	1	3.6

4.26 Units of Space Velocity

1/s	1/min	1/h	1/day
1	60	3600	86400
0.016667	1	60	1440
2.778×10^{-4}	0.016667	1	24
1.157×10^{-5}	6.944×10^{-4}	0.041667	1

4.19) $WHSV = \dfrac{\text{mass of feed per hour}}{\text{mass of catalyst}}$, 1/h

4.20) $LHSV = \dfrac{\text{volume of liquid feed per hour}}{\text{volume of catalyst}}$, 1/h

4.21) $GHSV = \dfrac{\text{volume of gaseous feed per hour}}{\text{volume of catalyst}}$, 1/h

4.27 Units of Temperature

	Temperature scale	Equation
4.20	Fahrenheit	°F = 1.8°C + 32
4.21	Celsius	°C = (°F − 32)/1.8
4.22	Kelvin	K = °C + 273.15
4.23	Rankine	°R = °F + 459.67
4.24	Réaumur	°Ré = 0.8°C

1 Δ°C = 1.8 Δ°F   1 ΔK = 1.8 Δ°R
1 Δ°R = 1 Δ°F   1 Δ°C = 1 ΔK

4.28 Units of Thermal Conductivity*

Btu(IT)/(h·ft·°F)	W/(m·°C)	kW/(m·°C)	cal(TC)/(s·cm·°C)
1	1.7307	0.00173073	0.0041366
0.57779	1	0.001	0.0023901
577.79	1000	1	2.3901
241.75	418.4	0.4184	1

*°C can be replaced with K and °F can be replaced with °R without affecting the value.

1 W/(cm·°C) = 100 W/(m·°C)   1 Btu(IT)·in/(h·ft^2·°F) = 0.14423 W/(m·°C)
1 cal(IT)/(s·cm·°C) = 418.68 W/(m·°C)

4.29 Units of Time

s	min	h	day
1	0.016667	2.778×10^{-4}	1.157×10^{-5}
60	1	0.016667	6.944×10^{-4}
3600	60	1	0.041667
86400	1440	24	1

1 year = 365 day
1 stream day = working hours in a day (depending on working days in a year)

4.30 Units of Velocity, Linear

in/s	ft/s	ft/min	ft/h	cm/s	m/s	m/h
1	0.083333	5	300	2.54	0.0254	91.44
12	1	60	3600	30.48	0.3048	1097.3
0.2	0.016667	1	60	0.508	0.00508	18.288
0.0033333	2.778×10^{-4}	0.016667	1	0.0084667	8.467×10^{-5}	0.3048
0.39370	0.032808	1.9685	118.11	1	0.01	36
39.370	3.2808	196.85	11811.0	100	1	3600
0.010936	9.113×10^{-4}	0.054681	3.2808	0.027778	2.778×10^{-4}	1

1 m/min = 0.016667 m/s   1 km/h = 0.27778 m/s
1 mile/h = 0.44704 m/s

4.31 Units of Viscosity (Absolute or Dynamic)

$lb_m/(ft \cdot h)$	$lb_f \cdot s/ft^2$	$g/(cm \cdot s)$ = Poise	cP = mPa·s	kg/(m·s)	Pa·s = N·s/m^2
1	8.634×10^{-6}	0.0041338	0.41338	4.134×10^{-4}	4.134×10^{-4}
115826.6	1	478.80	47880.3	47.880	47.880
241.91	0.0020885	1	100	0.1	0.1
2.4191	2.089×10^{-5}	0.01	1	0.001	0.001
2419.1	0.020885	10	1000	1	1

1 $lb_m/(ft \cdot s)$ = 1.4882 Pa·s   1 $lb_m/(in \cdot s)$ = 17.858 Pa·s
1 mN·s/m^2 = 1mPa·s = 0.001 Pa·s   1 kg/(m·h) = 2.778×10^{-4} Pa·s
1 dyn·s/cm^2 = 1 Poise = 0.1 Pa·s   1 $kg_f \cdot s/m^2$ = 9.8067 Pa·s
1 µP = 1×10^{-7} Pa·s

4.32 Units of Volume

in^3	ft^3	cm^3 = mL	m^3	L = dm^3	gal(US)
1	5.787×10^{-4}	16.387	1.639×10^{-5}	0.016387	0.0043290
1728	1	28316.8	0.028317	28.317	7.4805
0.061024	3.531×10^{-5}	1	1×10^{-6}	0.001	2.642×10^{-4}
61023.7	35.315	1×10^6	1	1000	264.17
61.024	0.035315	1000	0.001	1	0.26417
231	0.13368	3785.4	0.0037854	3.7854	1

1 cc = 1 cm^3
1 gal(imp) = 4.5461 L
1 fl oz(US) = 29.574 mL

1 mm^3 = 1×10^{-6} L
1 bbl(oil) = 42 gal(US)
1 pint(US) = 0.47318 L

4.33 Units of Volumetric Flowrate

ft^3/h	ft^3/s	m^3/s	m^3/h	L/min	gal(US)/min
1	2.778×10^{-4}	7.866×10^{-6}	0.028317	0.47195	0.12468
3600	1	0.028317	101.94	1699.0	448.83
127132.8	35.315	1	3600	60000	15850.3
35.315	0.0098096	2.778×10^{-4}	1	16.667	4.4029
2.1189	5.886×10^{-4}	1.667×10^{-5}	0.06	1	0.26417
8.0208	0.0022280	6.309×10^{-5}	0.22712	3.7854	1

1 cm^3/s = 1×10^{-6} m^3/s
1 μL/min = 1.667×10^{-11} m^3/s
1 in^3/s = 5.787×10^{-4} ft^3/s
1 gpm = 1 gal/min
1 MMSCFD = 1×10^6 standard ft^3/day

1 mL/min = 1.667×10^{-8} m^3/s
1 ft^3/min = 0.016667 ft^3/s
1 cusec = 1 ft^3/s
1 cfm = 1 ft^3/min

5.0 Indices

5.1) $a^0 = 1$

5.2) $a^{-n} = \dfrac{1}{a^n}$

5.3) $a^n = a^m \Rightarrow n = m$

5.4) $a^m \times a^n = a^{m+n}$

5.5) $\dfrac{a^m}{a^n} = a^{m-n}$

5.6) $(a^m)^n = a^{mn}$

5.7) $\sqrt[n]{a} = a^{\frac{1}{n}}$

5.8) $a^m \times b^m = (ab)^m$

5.9) $\dfrac{a^m}{b^m} = \left(\dfrac{a}{b}\right)^m$

5.10) $a^{\frac{m}{n}} = (a^{\frac{1}{n}})^m = (a^m)^{\frac{1}{n}}$

5.11) $(\sqrt[n]{a})^m = \sqrt[n]{a^m} = (a^m)^{\frac{1}{n}} = a^{\frac{m}{n}}$

5.12) $\sqrt[n]{ab} = (ab)^{\frac{1}{n}} = a^{\frac{1}{n}} b^{\frac{1}{n}}$

5.13) $\sqrt[n]{a+b} \neq \sqrt[n]{a} + \sqrt[n]{b}$

5.14) $\sqrt[n]{\dfrac{a}{b}} = \dfrac{\sqrt[n]{a}}{\sqrt[n]{b}} = \dfrac{a^{\frac{1}{n}}}{b^{\frac{1}{n}}} = \left(\dfrac{a}{b}\right)^{\frac{1}{n}}$

5.15) $\sqrt[m]{\sqrt[n]{a}} = (\sqrt[n]{a})^{\frac{1}{m}} = (a)^{\frac{1}{mn}} = \sqrt[nm]{a}$

6.0 Logarithms

6.1) $\log_a 1 = 0$

6.2) $\log_a a = 1$

6.3) $\log_{10} 10 = 1$

6.4) $\ln e = 1$ ($e = 2.7183$)

6.5) $\log_a\left(\dfrac{1}{x}\right) = -\log_a(x)$

6.6) $\log_a xy = \log_a x + \log_a y$

6.7) $\log_a\left(\dfrac{x}{y}\right) = \log_a x - \log_a y$

6.8) $\log_a x^n = n \log_a x$

6.9) $\log_a \sqrt[y]{x} = \dfrac{1}{y} \log_a x$

6.10) $\log_a x = \log_a y \Rightarrow x = y$

6.11) $\log_a b = \dfrac{\log_c b}{\log_c a}$

6.12) $\log_a b = \dfrac{1}{\log_b a}$

6.13) $\log_{10} e = \dfrac{1}{\ln 10}$

6.14) $y = a^x \Rightarrow x = \log_a y$

6.15) $y = 10^x \Rightarrow x = \log y$ (log is $\log_{10}$)

6.16) $y = e^x \Rightarrow x = \ln y$

7.0 Facts from Algebra

7.1) $(a+b)^2 = a^2 + b^2 + 2ab$ 7.2) $(a-b)^2 = a^2 + b^2 - 2ab$

7.3) $(a+b)^3 = a^3 + b^3 + 3ab(a+b) = a^3 + b^3 + 3a^2b + 3ab^2$

7.4) $(a-b)^3 = a^3 - b^3 - 3ab(a-b) = a^3 - b^3 - 3a^2b + 3ab^2$

7.5) $(a+b)^n = $ See Eq. 7.17

7.6) $(a+b+c)^2 = a^2 + b^2 + c^2 + 2ab + 2bc + 2ca$

7.7) $(a-b+c)^2 = a^2 + b^2 + c^2 - 2ab - 2bc + 2ca$

7.8) $(a+b+c)^3 = a^3 + b^3 + c^3 + 3a^2(b+c) + 3b^2(a+c) + 3c^2(a+b) + 6abc$

7.9) $(a+b+c+d)^2 = a^2 + b^2 + c^2 + d^2 + 2a(b+c+d) + 2b(c+d) + 2cd$

7.10) $a^2 - b^2 = (a+b)(a-b)$ 7.11) $a^2 + b^2 = (a+b\sqrt{-1})(a-b\sqrt{-1})$

7.12) $a^3 - b^3 = (a-b)(a^2 + ab + b^2)$ 7.13) $a^3 + b^3 = (a+b)(a^2 - ab + b^2)$

7.1 Binomial Theorem

For any real numbers a and b:

7.14) $(a+b)^n = \binom{n}{0}a^n b^0 + \binom{n}{1}a^{n-1}b^1 + \binom{n}{2}a^{n-2}b^2 + \cdots + \binom{n}{r}a^{n-r}b^r + \cdots + \binom{n}{n}a^0 b^n$

7.15) $\binom{n}{r} = \dfrac{n!}{r!(n-r)!} = \dfrac{n(n-1)(n-2)(n-3)\cdots(n-r+1)}{r!}$

7.16) $n! = n(n-1)(n-2)\times\cdots\times 3\times 2\times 1 = 1\times 2\times 3\times\cdots\times n$ (Remember $0! = 1$)

7.17) $(a+b)^n = a^n + na^{n-1}b + \dfrac{n(n-1)}{2!}a^{n-2}b^2 + \dfrac{n(n-1)(n-2)}{3!}a^{n-3}b^3 + \cdots + b^n$

7.2 Permutations and Combinations

7.2.1 Permutations

7.18) ${}^nP_r = \dfrac{n!}{(n-r)!} = n(n-1)(n-2)(n-3)\cdots(n-r+1)$

where ${}^nP_r$ denotes number of permutations of n different items taken r at a time. See Eqs. 7.15 and 7.16.

7.2.2 Combinations

7.19) $^nC_r = \dfrac{n!}{r!(n-r)!} = \dfrac{^nP_r}{r!} = \dfrac{n(n-1)(n-2)(n-3)\cdots(n-r+1)}{r!}$

where ^nC_r denotes number of combinations of n different things taken r at a time. See Eqs. 7.15 and 7.16.

7.20) $^nC_r = {^nC_{n-r}}$

7.3 Quadratic Equation

An equation of the second degree polynomial is called quadratic equation and the general form of a quadratic equation is

7.21) $ax^2 + bx + c = 0$

General quadratic formula or solution of the quadratic equation is

7.22) $x = \dfrac{-b \pm \sqrt{b^2 - 4ac}}{2a}$

If α and β are the two roots of the quadratic equation, then

7.23) $\alpha = \dfrac{-b + \sqrt{b^2 - 4ac}}{2a}$ 　　　　7.24) $\beta = \dfrac{-b - \sqrt{b^2 - 4ac}}{2a}$

Sum of the roots α and β can be written as

7.25) $S = \alpha + \beta = -\dfrac{b}{a}$

Product of the roots α and β can be written as

7.26) $P = \alpha \times \beta = \dfrac{c}{a}$

In terms of S and P, the quadratic equation can be written as

7.27) $x^2 - Sx + P = 0$

Discriminant ($Disc$) of the quadratic equation (Eq. 7.21) is

7.28) $Disc = b^2 - 4ac$

7.29) If $Disc = b^2 - 4ac = 0$ then roots are real and equal
7.30) If $Disc = b^2 - 4ac > 0$ then roots are real and unequal
7.31) If $Disc = b^2 - 4ac < 0$ then roots are complex conjugate

7.4 Cubic Equation

An equation of the third degree polynomial is called cubic equation and the general form of a cubic equation is

7.32) $x^3 + a_1 x^2 + a_2 x + a_3 = 0$

Solution of the above cubic equation is

7.33) $x_1 = C + D - \frac{1}{3} a_1$

7.34) $x_2 = -\frac{1}{2}(C + D) - \frac{1}{3} a_1 + \frac{1}{2} i\sqrt{3}(C - D)$

7.35) $x_3 = -\frac{1}{2}(C + D) - \frac{1}{3} a_1 - \frac{1}{2} i\sqrt{3}(C - D)$

7.36) $C = (B + \sqrt{A^3 + B^2})^{\frac{1}{3}}$ 7.37) $D = (B - \sqrt{A^3 + B^2})^{\frac{1}{3}}$

7.38) $A = \frac{3a_2 - a_1^2}{9}$ 7.39) $B = \frac{9 a_1 a_2 - 27 a_3 - 2 a_1^3}{54}$

7.40) $x_1 + x_2 + x_3 = -a_1$ 7.41) $x_1 x_2 + x_2 x_3 + x_3 x_1 = a_2$

7.42) $x_1 x_2 x_3 = -a_3$

Discriminant (*Disc*) of the cubic equation (Eq. 7.32) is

7.43) $Disc = A^3 + B^2$ (See Eqs. 7.38 and 7.39)

7.44) If $Disc = 0$ then all roots are real and at least two are equal
7.45) If $Disc > 0$ then one root is real and two complex conjugate
7.46) If $Disc < 0$ then all roots are real and unequal.

8.0 Trigonometry

8.1) $\sin\theta = \dfrac{y}{r}$

8.2) $\csc\theta = \dfrac{r}{y} = \dfrac{1}{\sin\theta}$

8.3) $\cos\theta = \dfrac{x}{r}$

8.4) $\sec\theta = \dfrac{r}{x} = \dfrac{1}{\cos\theta}$

8.5) $\tan\theta = \dfrac{y}{x}$

8.6) $\cot\theta = \dfrac{x}{y} = \dfrac{1}{\tan\theta}$

8.7) $\tan\theta = \dfrac{\sin\theta}{\cos\theta}$

8.8) $\cot\theta = \dfrac{\cos\theta}{\sin\theta}$

8.9) $\sin(-\theta) = -\sin(\theta)$

810) $\cos(-\theta) = \cos(\theta)$

8.11) $\tan(-\theta) = -\tan(\theta)$

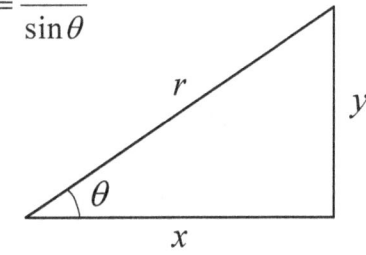

8.12) $\sin(\theta) = 2\sin\dfrac{\theta}{2}\cos\dfrac{\theta}{2} = \sqrt{\dfrac{1-\cos 2\theta}{2}}$

8.13) $\cos(\theta) = \cos^2\dfrac{\theta}{2} - \sin^2\dfrac{\theta}{2} = 2\cos^2\dfrac{\theta}{2} - 1 = 1 - 2\sin^2\dfrac{\theta}{2} = \sqrt{\dfrac{1+\cos 2\theta}{2}}$

8.14) $\sin(90° - \theta) = \cos\theta$

8.15) $\sin(90° + \theta) = \cos\theta$

8.16) $\cos(90° - \theta) = \sin\theta$

8.17) $\cos(90° + \theta) = -\sin\theta$

8.18) $\tan(90° - \theta) = \cot\theta$

8.19) $\tan(90° + \theta) = -\cot\theta$

8.20) $\sin(180° - \theta) = \sin\theta$

8.21) $\sin(180° + \theta) = -\sin\theta$

8.22) $\cos(180° - \theta) = -\cos\theta$

8.23) $\cos(180° + \theta) = -\cos\theta$

8.24) $\tan(180° - \theta) = -\tan\theta$

8.25) $\tan(180° + \theta) = \tan\theta$

8.26) $\sin^2\theta + \cos^2\theta = 1$

8.27) $1 + \tan^2\theta = \sec^2\theta$

8.28) $1 + \cot^2\theta = \csc^2\theta$

8.29) $\cos(\alpha - \beta) = \cos\alpha\cos\beta + \sin\alpha\sin\beta$

8.30) $\cos(\alpha + \beta) = \cos\alpha\cos\beta - \sin\alpha\sin\beta$

8.31) $\sin(\alpha - \beta) = \sin\alpha\cos\beta - \cos\alpha\sin\beta$

8.32) $\sin(\alpha + \beta) = \sin\alpha\cos\beta + \cos\alpha\sin\beta$

8.33) $\tan(\alpha + \beta) = \dfrac{\tan\alpha + \tan\beta}{1 - \tan\alpha\tan\beta}$

8.34) $\tan(\alpha - \beta) = \dfrac{\tan\alpha - \tan\beta}{1 + \tan\alpha\tan\beta}$

8.35) $\sin 2\alpha = 2\sin\alpha\cos\alpha = \dfrac{2\tan\alpha}{1+\tan^2\alpha}$

8.36) $\cos 2\alpha = \cos^2\alpha - \sin^2\alpha = 2\cos^2\alpha - 1 = 1 - 2\sin^2\alpha = \dfrac{1-\tan^2\alpha}{1+\tan^2\alpha}$

8.37) $\tan 2\alpha = \dfrac{2\tan\alpha}{1-\tan^2\alpha}$ 	8.38) $\sin\dfrac{\theta}{2} = \sqrt{\dfrac{1-\cos\theta}{2}}$

8.39) $\cos\dfrac{\theta}{2} = \sqrt{\dfrac{1+\cos\theta}{2}}$

8.40) $\tan\dfrac{\theta}{2} = \sqrt{\dfrac{1-\cos\theta}{1+\cos\theta}} = \dfrac{1-\cos\theta}{\sin\theta} = \dfrac{\sin\theta}{1+\cos\theta}$

8.41) $\sin\alpha + \sin\beta = 2\sin\dfrac{\alpha+\beta}{2}\cos\dfrac{\alpha-\beta}{2}$

8.42) $\sin\alpha - \sin\beta = 2\cos\dfrac{\alpha+\beta}{2}\sin\dfrac{\alpha-\beta}{2}$

8.43) $\cos\alpha + \cos\beta = 2\cos\dfrac{\alpha+\beta}{2}\cos\dfrac{\alpha-\beta}{2}$

8.44) $\cos\alpha - \cos\beta = -2\sin\dfrac{\alpha+\beta}{2}\sin\dfrac{\alpha-\beta}{2}$

8.45) $\tan\alpha + \tan\beta = \dfrac{\sin(\alpha+\beta)}{\cos\alpha\cos\beta}$ 	8.46) $\tan\alpha - \tan\beta = \dfrac{\sin(\alpha-\beta)}{\cos\alpha\cos\beta}$

8.47) $\sin\alpha\cos\beta = \dfrac{1}{2}\sin(\alpha+\beta) + \dfrac{1}{2}\sin(\alpha-\beta)$

8.48) $\cos\alpha\cos\beta = \dfrac{1}{2}\cos(\alpha+\beta) + \dfrac{1}{2}\cos(\alpha-\beta)$

8.49) $\sin\alpha\sin\beta = \dfrac{1}{2}\cos(\alpha-\beta) - \dfrac{1}{2}\cos(\alpha+\beta)$

8.50) $\tan\alpha\tan\beta = \dfrac{\tan\alpha+\tan\beta}{\cot\alpha+\cot\beta}$

8.1 Solution of a Triangle

Solution of a triangle means to find out the values of all its sides and angles, i.e., the length of all the three sides and measure of all the three angles.

8.51) $\alpha + \beta + \gamma = 180°$

8.1.1 Law of Sines

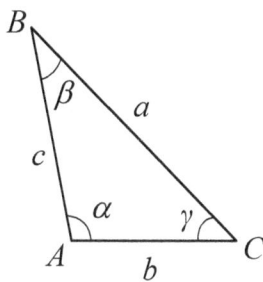

8.52) $\dfrac{a}{\sin\alpha} = \dfrac{b}{\sin\beta} = \dfrac{c}{\sin\gamma}$

8.53) $\dfrac{a}{\sin\alpha} = \dfrac{b}{\sin\beta}$

8.54) $\dfrac{b}{\sin\beta} = \dfrac{c}{\sin\gamma}$

8.55) $\dfrac{c}{\sin\gamma} = \dfrac{a}{\sin\alpha}$

8.1.2 Law of Cosines

8.56) $a^2 = b^2 + c^2 - 2bc\cos\alpha$

8.57) $\cos\alpha = \dfrac{b^2 + c^2 - a^2}{2bc}$

8.58) $b^2 = c^2 + a^2 - 2ac\cos\beta$

8.59) $\cos\beta = \dfrac{c^2 + a^2 - b^2}{2ac}$

8.60) $c^2 = a^2 + b^2 - 2ab\cos\gamma$

8.61) $\cos\gamma = \dfrac{a^2 + b^2 - c^2}{2ab}$

8.1.3 Law of Tangents

8.62) $\dfrac{a+b}{a-b} = \dfrac{\tan\tfrac{1}{2}(\alpha+\beta)}{\tan\tfrac{1}{2}(\alpha-\beta)}$

8.63) $\dfrac{b+c}{b-c} = \dfrac{\tan\tfrac{1}{2}(\beta+\gamma)}{\tan\tfrac{1}{2}(\beta-\gamma)}$

8.64) $\dfrac{c+a}{c-a} = \dfrac{\tan\tfrac{1}{2}(\gamma+\alpha)}{\tan\tfrac{1}{2}(\gamma-\alpha)}$

8.1.4 Half Angle Formulas

8.65) $s = \dfrac{1}{2}(a+b+c)$

8.66) $\sin\dfrac{\alpha}{2} = \sqrt{\dfrac{(s-b)(s-c)}{bc}}$

8.67) $\sin\dfrac{\beta}{2} = \sqrt{\dfrac{(s-c)(s-a)}{ca}}$

8.68) $\sin\dfrac{\gamma}{2} = \sqrt{\dfrac{(s-a)(s-b)}{ab}}$

8.69) $\cos\dfrac{\alpha}{2} = \sqrt{\dfrac{s(s-a)}{bc}}$

8.70) $\cos\dfrac{\beta}{2} = \sqrt{\dfrac{s(s-b)}{ca}}$

8.71) $\cos\dfrac{\gamma}{2} = \sqrt{\dfrac{s(s-c)}{ab}}$

8.72) $\tan\dfrac{\alpha}{2} = \sqrt{\dfrac{(s-b)(s-c)}{s(s-a)}}$

8.73) $\tan\dfrac{\beta}{2} = \sqrt{\dfrac{(s-c)(s-a)}{s(s-b)}}$

8.74) $\tan\dfrac{\gamma}{2} = \sqrt{\dfrac{(s-a)(s-b)}{s(s-c)}}$

8.2 Solution of a Right-Angled Triangle

A right-angled triangle is one which has one 90° angle in its dimensions.

8.51) $\alpha + \beta + 90° = 180°$

8.75) $c^2 = a^2 + b^2$ (Pythagoras Theorem)

8.76) $\sin\alpha = \dfrac{a}{c}$

8.77) $\sin\beta = \dfrac{b}{c}$

8.78) $\cos\alpha = \dfrac{b}{c}$

8.79) $\cos\beta = \dfrac{a}{c}$

8.80) $\tan\alpha = \dfrac{a}{b}$

8.81) $\tan\beta = \dfrac{b}{a}$

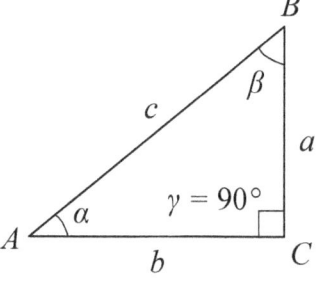

8.3 Area of a Triangle

If two sides and an angle between these two sides are known then area (Δ) is

8.82) $\Delta = \dfrac{1}{2}bc\sin\alpha$

8.83) $\Delta = \dfrac{1}{2}ca\sin\beta$

8.84) $\Delta = \dfrac{1}{2}ab\sin\gamma$

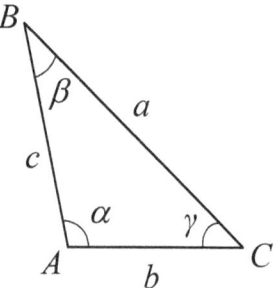

If one side and any two angles are known then area (Δ) is

8.85) $\Delta = \dfrac{1}{2}b^2\dfrac{\sin\gamma\sin\alpha}{\sin\beta}$

8.86) $\Delta = \dfrac{1}{2}a^2\dfrac{\sin\beta\sin\gamma}{\sin\alpha}$

8.87) $\Delta = \dfrac{1}{2}c^2 \dfrac{\sin\alpha \sin\beta}{\sin\gamma}$ 8.51) $\alpha + \beta + \gamma = 180°$

If all the three sides are known, then area (Δ) is

8.65) $s = \dfrac{1}{2}(a+b+c)$

8.88) $\Delta = \sqrt{s(s-a)(s-b)(s-c)}$ (Heron's formula)

See also Section 12.6.

8.4 Circumradius of a Triangle

The radius R of a circle (circumscribed or circumcircle) that passes through the vertices of a triangle is

8.89) $R = \dfrac{abc}{4\Delta} = \dfrac{abc}{4\sqrt{s(s-a)(s-b)(s-c)}}$

where Δ = area of the triangle.

8.65) $s = \dfrac{1}{2}(a+b+c)$

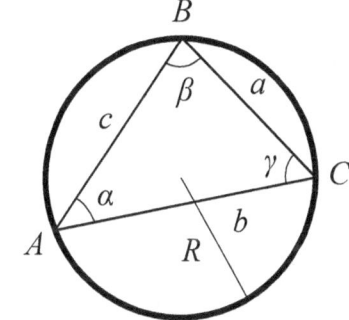

8.5 Inradius of a Triangle

The radius r of a circle (inscribed circle) which touches the sides of a triangle is

8.90) $r = \dfrac{\Delta}{s} = \dfrac{\sqrt{s(s-a)(s-b)(s-c)}}{s}$

where Δ = area of the triangle

8.65) $s = \dfrac{1}{2}(a+b+c)$

8.91) $r = 4R \sin\dfrac{\alpha}{2} \sin\dfrac{\beta}{2} \sin\dfrac{\gamma}{2}$

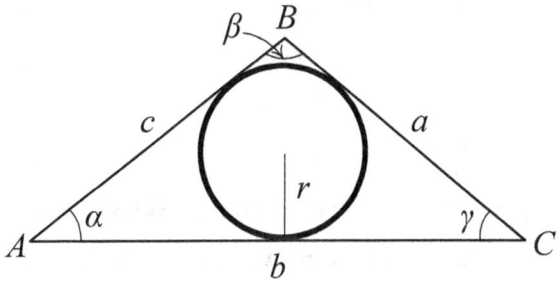

8.92) $r = (s-a)\tan\dfrac{\alpha}{2} = (s-b)\tan\dfrac{\beta}{2} = (s-c)\tan\dfrac{\gamma}{2}$

8.6 Inverse Trigonometric Functions

8.93) $\theta = \arcsin(x) = \sin^{-1} x \Rightarrow x = \sin\theta$

8.94) $\theta = \arccos(x) = \cos^{-1} x \Rightarrow x = \cos\theta$

8.95) $\theta = \arctan(x) = \tan^{-1} x \Rightarrow x = \tan\theta$

8.96) $\sin^{-1} x + \cos^{-1} x = \dfrac{\pi}{2}$

8.97) $\tan^{-1} x + \cot^{-1} x = \dfrac{\pi}{2}$

8.98) $\sec^{-1} x + \csc^{-1} x = \dfrac{\pi}{2}$

8.99) $\sin^{-1}(-x) = -\sin^{-1} x$

8.100) $\cos^{-1}(-x) = \pi - \cos^{-1} x$

8.101) $\tan^{-1}(-x) = -\tan^{-1} x$

8.7 Hyperbolic Functions

8.102) $\sinh x = \dfrac{e^x - e^{-x}}{2}$

8.103) $\cosh x = \dfrac{e^x + e^{-x}}{2}$

8.104) $\tanh x = \dfrac{e^x - e^{-x}}{e^x + e^{-x}}$

8.105) $\tanh x = \dfrac{\sinh x}{\cosh x}$

8.106) $\operatorname{csch} x = \dfrac{1}{\sinh x}$

8.107) $\operatorname{sech} x = \dfrac{1}{\cosh x}$

8.108) $\coth x = \dfrac{1}{\tanh x}$

8.109) $\coth x = \dfrac{\cosh x}{\sinh x}$

8.110) $\cosh^2 x - \sinh^2 x = 1$

8.111) $\operatorname{sech}^2 x + \tanh^2 x = 1$

8.112) $\coth^2 x - \operatorname{csch}^2 x = 1$

8.113) $\sinh(-x) = -\sinh x$

8.114) $\cosh(-x) = \cosh x$

8.115) $\tanh(-x) = -\tanh x$

8.116) $\sinh(x+y) = \sinh x \cosh y + \cosh x \sinh y$

8.117) $\sinh(x-y) = \sinh x \cosh y - \cosh x \sinh y$

8.118) $\cosh(x+y) = \cosh x \cosh y + \sinh x \sinh y$

8.119) $\cosh(x-y) = \cosh x \cosh y - \sinh x \sinh y$

8.120) $\tanh(x+y) = \dfrac{\tanh x + \tanh y}{1 + \tanh x \tanh y}$ 	8.121) $\tanh(x-y) = \dfrac{\tanh x - \tanh y}{1 - \tanh x \tanh y}$

8.122) $\sinh x + \sinh y = 2\sinh\left(\dfrac{x+y}{2}\right)\cosh\left(\dfrac{x-y}{2}\right)$

8.123) $\sinh x - \sinh y = 2\sinh\left(\dfrac{x-y}{2}\right)\cosh\left(\dfrac{x+y}{2}\right)$

8.124) $\cosh x + \cosh y = 2\cosh\left(\dfrac{x+y}{2}\right)\cosh\left(\dfrac{x-y}{2}\right)$

8.125) $\cosh x - \cosh y = 2\sinh\left(\dfrac{x+y}{2}\right)\sinh\left(\dfrac{x-y}{2}\right)$

8.126) $\tanh x + \tanh y = \dfrac{\sinh(x+y)}{\cosh x \cosh y}$ 	8.127) $\tanh x - \tanh y = \dfrac{\sinh(x-y)}{\cosh x \cosh y}$

8.8 Inverse Hyperbolic Functions

8.128) $\sinh^{-1} x = \ln(x + \sqrt{x^2 + 1})$ 	8.129) $\cosh^{-1} x = \ln(x + \sqrt{x^2 - 1})$

8.130) $\tanh^{-1} x = \dfrac{1}{2}\ln\dfrac{1+x}{1-x}$ 	8.131) $\sinh^{-1}(-x) = -\sinh^{-1} x$

8.132) $\tanh^{-1}(-x) = -\tanh^{-1} x$

9.0 Series

9.1 Arithmetic Series

General form of an arithmetic series is

9.1) $a+(a+d)+(a+2d)+\cdots+[a+(n-1)d]$

where a is the first term and d is the common difference between successive terms.

nth term (a_n) of an arithmetic series is

9.2) $a_n = a+(n-1)d$

Sum of n terms of an arithmetic series is

9.3) $S_n = \dfrac{n}{2}[2a+(n-1)d]$

Examples of arithmetic series are

9.4) $1+2+3+\cdots+n = \dfrac{n(n+1)}{2}$ 9.5) $1+3+5+\cdots+(2n-1)=n^2$

9.6) $4+8+12+\cdots+4n = 2n(n+1)$

9.2 Geometric Series

General form of a geometric series is

9.7) $a+ar+ar^2+\cdots+ar^{n-1}$

where a is the first term and r is the geometric ratio between any two successive terms, e.g., $r = \dfrac{ar^2}{ar}$.

nth term (a_n) of a geometric series is

9.8) $a_n = ar^{n-1}$

Sum of n terms of a geometric series is

9.9) $S_n = \dfrac{a(1-r^n)}{1-r}$

When geometric series is

9.10) $a + ar + ar^2 + \cdots \infty$, then sum of the series is

9.11) $S_\infty = \dfrac{a}{1-r}$

Examples of geometric series are

9.12) $3^1 + 3^2 + 3^3 + \cdots + 3^n = \dfrac{3}{2}(3^n - 1)$ 9.13) $5 + 10 + 20 + \cdots + n \, terms = 5(2^n - 1)$

9.14) $\dfrac{1}{8} + \dfrac{1}{16} + \dfrac{1}{32} + \cdots = \dfrac{1}{4}$

9.3 Binomial Series

9.15) $(1+x)^n = 1 + nx + \dfrac{n(n-1)}{2!}x^2 + \dfrac{n(n-1)(n-2)}{3!}x^3 + \cdots$

9.16) $(1+x)^{-1} = 1 - x + x^2 - x^3 + x^4 - \cdots$

9.17) $(1+x)^{-2} = 1 - 2x + 3x^2 - 4x^3 + 5x^4 - \cdots$

9.18) $(1+x)^{\frac{1}{2}} = 1 + \dfrac{1}{2}x - \dfrac{1}{8}x^2 + \dfrac{1}{16}x^3 - \cdots$ 9.19) $(1+x)^{-\frac{1}{2}} = 1 - \dfrac{1}{2}x + \dfrac{3}{8}x^2 - \dfrac{5}{16}x^3 + \cdots$

9.4 Taylor Series

9.20) $f(x) = f(a) + f'(a)(x-a) + \dfrac{f''(a)(x-a)^2}{2!} + \cdots + \dfrac{f^{(n-1)}(a)(x-a)^{n-1}}{(n-1)!} + R_n$

where R_n is the remainder after n terms.

If $a = 0$, Taylor series is known as Maclaurin series.

10.0 Matrices

A matrix of $m \times n$ order can be written as

$$10.1) \begin{pmatrix} a_{11} & a_{12} & a_{13} & \cdots & a_{1n} \\ a_{21} & a_{22} & a_{23} & \cdots & a_{2n} \\ a_{31} & a_{32} & a_{33} & \cdots & a_{3n} \\ \vdots & \vdots & \vdots & \ddots & \vdots \\ a_{m1} & a_{m2} & a_{m3} & \cdots & a_{mn} \end{pmatrix}$$

where m and n represent number of rows and number of columns, respectively.

Square matrix of 3×3 order is

$$10.2) \ A = \begin{pmatrix} a_{11} & a_{12} & a_{13} \\ a_{21} & a_{22} & a_{23} \\ a_{31} & a_{32} & a_{33} \end{pmatrix}$$

Identity matrix of 3×3 order is

$$10.3) \ I = \begin{pmatrix} 1 & 0 & 0 \\ 0 & 1 & 0 \\ 0 & 0 & 1 \end{pmatrix}$$

Null matrix of 3×3 order is

$$10.4) \ N = \begin{pmatrix} 0 & 0 & 0 \\ 0 & 0 & 0 \\ 0 & 0 & 0 \end{pmatrix}$$

10.1 Addition of Matrices A and B

Two matrices can be added only if their orders are same. The sum of two square matrices of 3×3 order is

10.5) $A + B = \begin{pmatrix} a_{11} & a_{12} & a_{13} \\ a_{21} & a_{22} & a_{23} \\ a_{31} & a_{32} & a_{33} \end{pmatrix} + \begin{pmatrix} b_{11} & b_{12} & b_{13} \\ b_{21} & b_{22} & b_{23} \\ b_{31} & b_{32} & b_{33} \end{pmatrix} = \begin{pmatrix} a_{11}+b_{11} & a_{12}+b_{12} & a_{13}+b_{13} \\ a_{21}+b_{21} & a_{22}+b_{22} & a_{23}+b_{23} \\ a_{31}+b_{31} & a_{32}+b_{32} & a_{33}+b_{33} \end{pmatrix}$

10.6) $A + B = B + A$

10.7) $A + (B + C) = (A + B) + C$

10.2 Subtraction of Matrices: Matrix B from Matrix A

Two matrices can be subtracted only if their orders are same. The subtraction of two square matrices of 3×3 order is

10.8) $A - B = \begin{pmatrix} a_{11} & a_{12} & a_{13} \\ a_{21} & a_{22} & a_{23} \\ a_{31} & a_{32} & a_{33} \end{pmatrix} - \begin{pmatrix} b_{11} & b_{12} & b_{13} \\ b_{21} & b_{22} & b_{23} \\ b_{31} & b_{32} & b_{33} \end{pmatrix} = \begin{pmatrix} a_{11}-b_{11} & a_{12}-b_{12} & a_{13}-b_{13} \\ a_{21}-b_{21} & a_{22}-b_{22} & a_{23}-b_{23} \\ a_{31}-b_{31} & a_{32}-b_{32} & a_{33}-b_{33} \end{pmatrix}$

10.9) $A - B = A + (-B)$

10.3 Multiplication of Matrix A by a Scalar

Multiplication of a matrix of 3×3 order by a scalar λ is

10.10) $\lambda A = \lambda \times \begin{pmatrix} a_{11} & a_{12} & a_{13} \\ a_{21} & a_{22} & a_{23} \\ a_{31} & a_{32} & a_{33} \end{pmatrix} = \begin{pmatrix} \lambda a_{11} & \lambda a_{12} & \lambda a_{13} \\ \lambda a_{21} & \lambda a_{22} & \lambda a_{23} \\ \lambda a_{31} & \lambda a_{32} & \lambda a_{33} \end{pmatrix}$

10.11) $\lambda(A + B) = \lambda A + \lambda B$

10.4 Multiplication of Matrix A and Matrix B

Two matrices can be multiplied only if the number of columns of the first matrix is equal to the number of rows of the second matrix. The final matrix is of the order of number of rows of the first matrix × number of columns of the second matrix.

10.12) $AB = \begin{pmatrix} a_{11} & a_{12} \\ a_{21} & a_{22} \\ a_{31} & a_{32} \end{pmatrix} \begin{pmatrix} b_{11} & b_{12} & b_{13} \\ b_{21} & b_{22} & b_{23} \end{pmatrix} = \begin{pmatrix} a_{11}b_{11}+a_{12}b_{21} & a_{11}b_{12}+a_{12}b_{22} & a_{11}b_{13}+a_{12}b_{23} \\ a_{21}b_{11}+a_{22}b_{21} & a_{21}b_{12}+a_{22}b_{22} & a_{21}b_{13}+a_{22}b_{23} \\ a_{31}b_{11}+a_{32}b_{21} & a_{31}b_{12}+a_{32}b_{22} & a_{31}b_{13}+a_{32}b_{23} \end{pmatrix}$

10.13) $AB \neq BA$ (not always true)

10.14) $A(B+C) = AB + AC$

10.15) $A(BC) = (AB)C$

If matrix A is multiplied by identity matrix I, the answer is the same as matrix A. The multiplication of a matrix of 3×3 order with identity matrix of 3×3 order is

10.16) $AI = \begin{pmatrix} a_{11} & a_{12} & a_{13} \\ a_{21} & a_{22} & a_{23} \\ a_{31} & a_{32} & a_{33} \end{pmatrix} \times \begin{pmatrix} 1 & 0 & 0 \\ 0 & 1 & 0 \\ 0 & 0 & 1 \end{pmatrix} = \begin{pmatrix} a_{11} & a_{12} & a_{13} \\ a_{21} & a_{22} & a_{23} \\ a_{31} & a_{32} & a_{33} \end{pmatrix} = A$

10.17) $AI = IA = A$

10.5 Determinant of Matrix A

Determinant of a matrix of 3×3 order is

10.18) $\det(A) = |A| = \begin{vmatrix} a_{11} & a_{12} & a_{13} \\ a_{21} & a_{22} & a_{23} \\ a_{31} & a_{32} & a_{33} \end{vmatrix} = a_{11} \begin{vmatrix} a_{22} & a_{23} \\ a_{32} & a_{33} \end{vmatrix} - a_{12} \begin{vmatrix} a_{21} & a_{23} \\ a_{31} & a_{33} \end{vmatrix} + a_{13} \begin{vmatrix} a_{21} & a_{22} \\ a_{31} & a_{32} \end{vmatrix}$

$= a_{11}(a_{22}a_{33} - a_{23}a_{32}) - a_{12}(a_{21}a_{33} - a_{23}a_{31}) + a_{13}(a_{21}a_{32} - a_{22}a_{31})$

10.6 Inverse of Matrix A

A^{-1} exists only and only if $\det(A)$ is not zero, i.e., A is not singular. See Section 10.5 and Section 10.8.

10.19) $AA^{-1} = I = A^{-1}A$

10.7 Transpose of Matrix A

The elements of the rows of matrix A become elements of the columns of the new matrix. The transpose (A^t) of matrix A of 3×3 order is

10.20) $A^t = \begin{pmatrix} a_{11} & a_{21} & a_{31} \\ a_{12} & a_{22} & a_{32} \\ a_{13} & a_{23} & a_{33} \end{pmatrix}$ of $A = \begin{pmatrix} a_{11} & a_{12} & a_{13} \\ a_{21} & a_{22} & a_{23} \\ a_{31} & a_{32} & a_{33} \end{pmatrix}$

10.8) Solution of Simultaneous Equations by Matrix Inverse Method

A set of three linear algebraic equations is

10.21) $\begin{aligned} a_{11}x_1 + a_{12}x_2 + a_{13}x_3 &= b_1 \\ a_{21}x_1 + a_{22}x_2 + a_{23}x_3 &= b_2 \\ a_{31}x_1 + a_{32}x_2 + a_{33}x_3 &= b_3 \end{aligned}$

Matrix formulation for the above three equations (Eq. 10.21) can be written as

10.22) $\begin{pmatrix} a_{11} & a_{12} & a_{13} \\ a_{21} & a_{22} & a_{23} \\ a_{31} & a_{32} & a_{33} \end{pmatrix} \begin{pmatrix} x_1 \\ x_2 \\ x_3 \end{pmatrix} = \begin{pmatrix} b_1 \\ b_2 \\ b_3 \end{pmatrix}$

10.23) $AX = B$

10.24) $X = A^{-1}B$

Matrix C consists of the cofactors of the elements of matrix A:

10.25) $C = \begin{pmatrix} \begin{vmatrix} a_{22} & a_{23} \\ a_{32} & a_{33} \end{vmatrix} & -\begin{vmatrix} a_{21} & a_{23} \\ a_{31} & a_{33} \end{vmatrix} & \begin{vmatrix} a_{21} & a_{22} \\ a_{31} & a_{32} \end{vmatrix} \\ -\begin{vmatrix} a_{12} & a_{13} \\ a_{32} & a_{33} \end{vmatrix} & \begin{vmatrix} a_{11} & a_{13} \\ a_{31} & a_{33} \end{vmatrix} & -\begin{vmatrix} a_{11} & a_{12} \\ a_{31} & a_{32} \end{vmatrix} \\ \begin{vmatrix} a_{12} & a_{13} \\ a_{22} & a_{23} \end{vmatrix} & -\begin{vmatrix} a_{11} & a_{13} \\ a_{21} & a_{23} \end{vmatrix} & \begin{vmatrix} a_{11} & a_{12} \\ a_{21} & a_{22} \end{vmatrix} \end{pmatrix}$

10.26) $A^{-1} = \dfrac{\operatorname{adj} A}{\det(A)} = \dfrac{C^t}{|A|} = \dfrac{\begin{pmatrix} \begin{vmatrix} a_{22} & a_{23} \\ a_{32} & a_{33} \end{vmatrix} & -\begin{vmatrix} a_{12} & a_{13} \\ a_{32} & a_{33} \end{vmatrix} & \begin{vmatrix} a_{12} & a_{13} \\ a_{22} & a_{23} \end{vmatrix} \\ -\begin{vmatrix} a_{21} & a_{23} \\ a_{31} & a_{33} \end{vmatrix} & \begin{vmatrix} a_{11} & a_{13} \\ a_{31} & a_{33} \end{vmatrix} & -\begin{vmatrix} a_{11} & a_{13} \\ a_{21} & a_{23} \end{vmatrix} \\ \begin{vmatrix} a_{21} & a_{22} \\ a_{31} & a_{32} \end{vmatrix} & -\begin{vmatrix} a_{11} & a_{12} \\ a_{31} & a_{32} \end{vmatrix} & \begin{vmatrix} a_{11} & a_{12} \\ a_{21} & a_{22} \end{vmatrix} \end{pmatrix}}{|A|}$

Eq. 10.24 then yields

10.27) $\begin{pmatrix} x_1 \\ x_2 \\ x_3 \end{pmatrix} = \frac{1}{|A|} \times \begin{pmatrix} \begin{vmatrix} a_{22} & a_{23} \\ a_{32} & a_{33} \end{vmatrix} & -\begin{vmatrix} a_{12} & a_{13} \\ a_{32} & a_{33} \end{vmatrix} & \begin{vmatrix} a_{12} & a_{13} \\ a_{22} & a_{23} \end{vmatrix} \\ -\begin{vmatrix} a_{21} & a_{23} \\ a_{31} & a_{33} \end{vmatrix} & \begin{vmatrix} a_{11} & a_{13} \\ a_{31} & a_{33} \end{vmatrix} & -\begin{vmatrix} a_{11} & a_{13} \\ a_{21} & a_{23} \end{vmatrix} \\ \begin{vmatrix} a_{21} & a_{22} \\ a_{31} & a_{32} \end{vmatrix} & -\begin{vmatrix} a_{11} & a_{12} \\ a_{31} & a_{32} \end{vmatrix} & \begin{vmatrix} a_{11} & a_{12} \\ a_{21} & a_{22} \end{vmatrix} \end{pmatrix} \times \begin{pmatrix} b_1 \\ b_2 \\ b_3 \end{pmatrix}$

$= \frac{1}{|A|} \times \begin{pmatrix} a_{22}a_{33} - a_{23}a_{32} & a_{13}a_{32} - a_{12}a_{33} & a_{12}a_{23} - a_{13}a_{22} \\ a_{23}a_{31} - a_{21}a_{33} & a_{11}a_{33} - a_{13}a_{31} & a_{13}a_{21} - a_{11}a_{23} \\ a_{21}a_{32} - a_{22}a_{31} & a_{12}a_{31} - a_{11}a_{32} & a_{11}a_{22} - a_{12}a_{21} \end{pmatrix} \times \begin{pmatrix} b_1 \\ b_2 \\ b_3 \end{pmatrix}$

Solving right hand side of Eq. 10.27 will give the values of x_1, x_2, and x_3.

10.9 Cramer Rule

A set of three linear algebraic equations is

10.21) $\begin{aligned} a_{11}x_1 + a_{12}x_2 + a_{13}x_3 &= b_1 \\ a_{21}x_1 + a_{22}x_2 + a_{23}x_3 &= b_2 \\ a_{31}x_1 + a_{32}x_2 + a_{33}x_3 &= b_3 \end{aligned}$

Matrix formulation for the above three equations (Eq. 10.21) can be written as

10.22) $\begin{pmatrix} a_{11} & a_{12} & a_{13} \\ a_{21} & a_{22} & a_{23} \\ a_{31} & a_{32} & a_{33} \end{pmatrix} \begin{pmatrix} x_1 \\ x_2 \\ x_3 \end{pmatrix} = \begin{pmatrix} b_1 \\ b_2 \\ b_3 \end{pmatrix}$

10.28) $x_1 = \dfrac{\begin{vmatrix} b_1 & a_{12} & a_{13} \\ b_2 & a_{22} & a_{23} \\ b_3 & a_{32} & a_{33} \end{vmatrix}}{\det(A)}$
10.29) $x_2 = \dfrac{\begin{vmatrix} a_{11} & b_1 & a_{13} \\ a_{21} & b_2 & a_{23} \\ a_{31} & b_3 & a_{33} \end{vmatrix}}{\det(A)}$

10.30) $x_3 = \dfrac{\begin{vmatrix} a_{11} & a_{12} & b_1 \\ a_{21} & a_{22} & b_2 \\ a_{31} & a_{32} & b_3 \end{vmatrix}}{\det(A)}$

10.10 Eigen Values and Eigen Vectors

For matrix A, an eigen value problem can be written as

10.31) $Av = \lambda v$

where A is $n \times n$ matrix, v is non-zero $n \times 1$ matrix, and λ is a scalar.

For non-zero v, one can write

10.32) $|A - \lambda I| = 0$

The solutions of Eq. 10.32, i.e., values of λ are called the eigen values of matrix A. For each eigen value there is an eigen vector.

11.0 Complex Numbers

A complex number is generally written as

11.1) $z = a + bi$

where a is real part while bi is imaginary part and a and b both are real numbers while i is imaginary unit.

11.2) $i = \sqrt{-1}$ 11.3) $i^2 = -1$

11.4) $i^3 = -i$ 11.5) $i^4 = 1$

11.6) $i^{-1} = -i$ 11.7) $i^{-2} = -1$

11.8) $i^{-3} = i$ 11.9) $i^{-4} = 1$

Conjugate of $a + bi$ is

11.10) $a - bi$ and vice-versa

When $z_1 = a_1 + b_1 i$ and $z_2 = a_2 + b_2 i$ then

11.11) $z_1 + z_2 = (a_1 + b_1 i) + (a_2 + b_2 i) = (a_1 + a_2) + (b_1 + b_2)i$

11.12) $z_1 - z_2 = (a_1 - a_2) + (b_1 - b_2)i$ 11.13) $z_1 \times z_2 = (a_1 a_2 - b_1 b_2) + (a_1 b_2 + a_2 b_1)i$

11.14) $\dfrac{z_1}{z_2} = \dfrac{a_1 a_2 + b_1 b_2}{a_2^2 + b_2^2} + \left(\dfrac{a_2 b_1 - a_1 b_2}{a_2^2 + b_2^2}\right)i$

11.15) $\sqrt{z} = \sqrt{a + bi} = \pm\left(\sqrt{\dfrac{a + \sqrt{a^2 + b^2}}{2}} + \sqrt{\dfrac{-a + \sqrt{a^2 + b^2}}{2}} \cdot i\right)$

Polar form a complex number $z = a + bi = x + yi$ (Cartesian form) is

11.16) $z = r(\cos\theta + i\sin\theta)$

where r is modulus and θ is argument or amplitude of z.

11.17) $r = \sqrt{a^2 + b^2}$

11.18) $\theta = \tan^{-1}\dfrac{b}{a}$

11.19) $z_1 z_2 = r_1 r_2 (\cos(\theta_1 + \theta_2) + i\sin(\theta_1 + \theta_2))$

11.20) $\dfrac{z_1}{z_2} = \dfrac{r_1}{r_2}(\cos(\theta_1 - \theta_2) + i\sin(\theta_1 - \theta_2))$

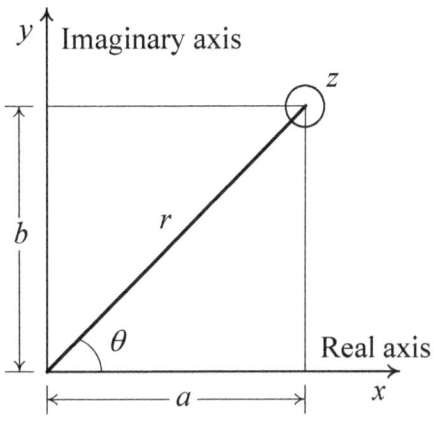

Exponential form of a complex number is

11.21) $z = re^{i\theta}$

For r and θ, see Eq. 11.17 and Eq. 11.18, respectively.

11.22) $e^{i\theta} = \cos\theta + i\sin\theta$

11.23) $e^{-i\theta} = \cos\theta - i\sin\theta$

De Moivre's theorem can be written as

11.24) $(r(\cos\theta + i\sin\theta))^n = r^n(\cos n\theta + i\sin n\theta)$

12.0 Plane Geometry

12.1 Square

12.1) *Area:* $A = a^2 = \dfrac{d^2}{2}$

12.2) *Perimeter:* $P = 4a$

12.3) $d = a\sqrt{2}$

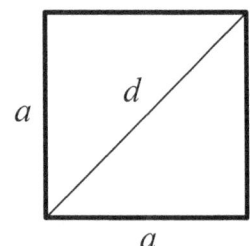

12.2 Rectangle

12.4) *Area:* $A = ab$

12.5) *Perimeter:* $P = 2a + 2b = 2(a+b)$

12.6) $d = \sqrt{a^2 + b^2}$

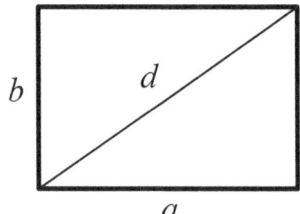

12.3 Rhombus

12.7) *Area:* $A = ah = \dfrac{1}{2}d_1 d_2 = a^2 \sin\alpha$

12.8) *Perimeter:* $P = 4a$

See also parallelogram.

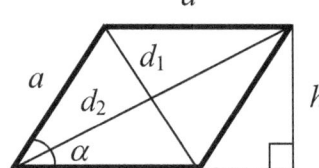

12.4 Parallelogram

12.9) $\alpha + \beta = 180°$

12.10) $h = b\sin\alpha$

12.11) *Area:* $A = ah = ab\sin\alpha$

12.12) *Area:* $A = \dfrac{1}{2}d_1 d_2 \sin\theta$

12.13) *Perimeter:* $P = 2a + 2b = 2(a+b)$

12.14) $d_1 = \sqrt{a^2 + b^2 - 2ab\cos\alpha}$ 12.15) $d_2 = \sqrt{a^2 + b^2 + 2ab\cos\alpha}$

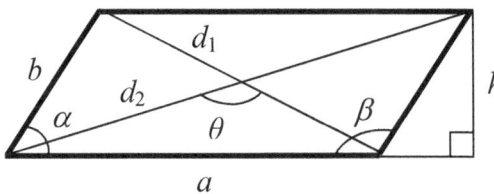

12.5 Trapezoid

12.16) $m = \dfrac{a+b}{2}$

12.17) *Area:* $A = \dfrac{h}{2}(a+b) = mh$

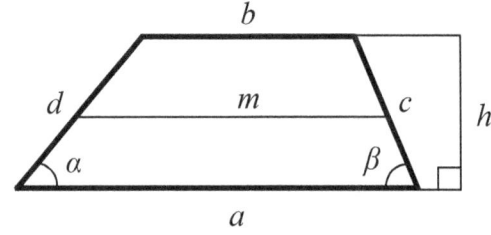

12.18) *Perimeter*: $P = a+b+c+d = a+b+h\left(\dfrac{1}{\sin\alpha}+\dfrac{1}{\sin\beta}\right)$

12.6 Triangle

For all types of triangles

12.19) *Area* $= \dfrac{1}{2} \times base \times height$

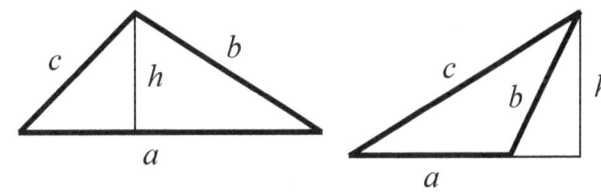

12.20) *Area*: $A = \dfrac{1}{2}ah$

12.21) *Perimeter*: $P = a+b+c$ See also Section 8.2 and Section 8.3.

12.6.1 Right-Angled Triangle

One of the three angles is of 90°.

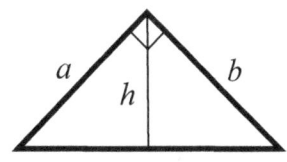

12.22) *Area*: $A = \dfrac{1}{2}ab = \dfrac{1}{2}ch$

12.23) *Perimeter*: $P = a+b+c$

12.24) $c = \sqrt{a^2+b^2}$ (Pythagoras Theorem)

12.25) $h = \dfrac{ab}{c}$

12.6.2 Equilateral Triangle

12.26) *Area*: $A = \dfrac{a^2\sqrt{3}}{4}$

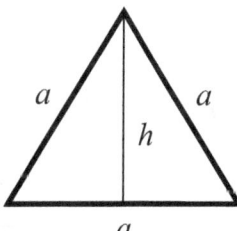

12.27) $h = \dfrac{a\sqrt{3}}{2}$

12.28) *Perimeter*: $P = 3a$

See also Section 8.2 and Section 8.3.

12.7 Polygon

12.7.1 Polygon, Irregular

Divide the polygon into convenient number of basic shapes (usually triangles). Measure the area of each shape and add these to find the total area of the polygon.

12.29) *Area*: $A = A_1 + A_2 + \cdots + A_n$

where n = number of the shapes (triangles) in which polygon is divided.

12.7.1.1 Quadrilateral, Irregular

Divide the irregular quadrilateral into two or more convenient triangles (or any other basic shapes) and measure the area of each triangle and add the areas.

For the figure shown

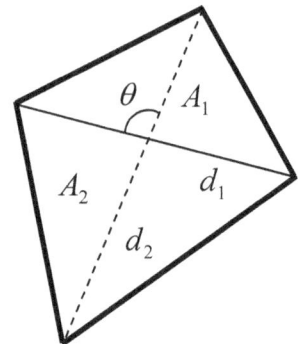

12.30) *Area*: $A = A_1 + A_2$

12.31) *Area*: $A = \frac{1}{2} d_1 d_2 \sin\theta$

12.7.2 Polygon, Regular

12.32) *Area*: $A = \frac{1}{2} \times perimeter \times apothem = \frac{1}{2} nah$

where n = number of sides of the polygon (e.g., $n = 5$ for pentagon).

12.33) *Area*: $A = \frac{1}{2} nr^2 \sin\left(\frac{360°}{n}\right)$

12.34) *Area*: $A = \frac{1}{4} na^2 \cot\left(\frac{180°}{n}\right)$

12.35) $\alpha = \frac{n-2}{n} \times 180°$

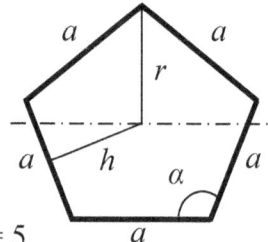

Pentagon, $n = 5$

12.36) *Perimeter*: $P = na$

12.37) *Sum of all angles* = $\sum \alpha = (n-2) \times 180°$ (e.g., for square it is 360°)

12.38) $a = 2r \sin\frac{180°}{n}$

12.39) $r = \dfrac{a}{2\sin\left(\dfrac{180°}{n}\right)}$

12.8 Circle

12.40) *Area*: $A = \dfrac{\pi d^2}{4} = \pi r^2$

12.41) *Perimeter*: $P = \pi d = 2\pi r$

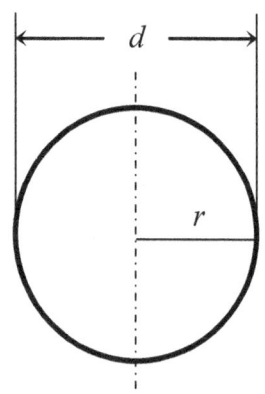

42

12.9 Annulus

12.42) Area: $A = \dfrac{\pi(D^2 - d^2)}{4} = \pi(R^2 - r^2)$

12.43) $b = \dfrac{D-d}{2} = R - r$

12.44) Perimeter: $P = \pi(D + d) = 2\pi(R + r)$

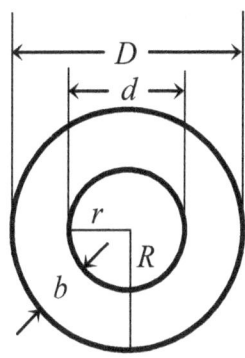

12.10 Sector of a Circle

12.45) Area: $A = \dfrac{1}{2} rs$

12.46) Area: $A = \dfrac{1}{2} r^2 \theta$ (θ in radian)

12.47) Area: $A = \dfrac{\pi r^2 \theta}{360°}$ (θ in degree)

12.48) $s = r\theta$ (θ in radian)

12.49) $s = r \dfrac{\pi \theta}{180°}$ (θ in degree)

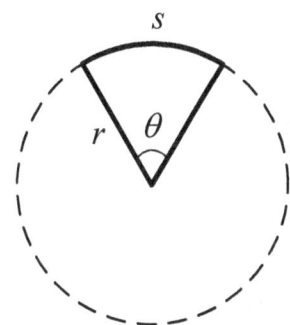

12.11 Sector of an Annulus

12.50) Area: $A = \dfrac{1}{2}(Rs_1 - rs_2)$

12.51) Area: $A = \dfrac{1}{2} \theta(R^2 - r^2)$ (θ in radian)

12.52) Area: $A = \dfrac{\pi \theta (R^2 - r^2)}{360°}$ (θ in degree)

12.53) Perimeter: $P = s_1 + s_2 + 2b$

12.54) $b = R - r$

12.55) $s_1 = R\theta$ (θ in radian)

12.56) $s_2 = r\theta$ (θ in radian)

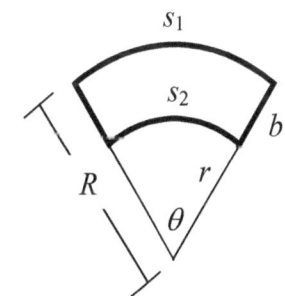

12.12 Segment of a Circle

12.57) *Area*: $A = \dfrac{r^2}{2}(\theta - \sin\theta)$ (θ in radian)

12.58) *Area*: $A = \dfrac{r^2}{2}\left(\dfrac{\pi\theta}{180°} - \sin\theta\right)$ (θ in degree)

12.59) $b = 2r\sin\dfrac{\theta}{2}$

12.60) $h = r\left(1 - \cos\dfrac{\theta}{2}\right)$

12.61) *Perimeter*: $P = s + b$

12.62) $s = r\theta$ (θ in radian)   12.63) $s = r\dfrac{\pi\theta}{180°}$ (θ in degree)

12.13 Ellipse

12.64) *Area*: $A = \pi ab$

12.65) *Area*: $A = \dfrac{\pi}{4}Dd$

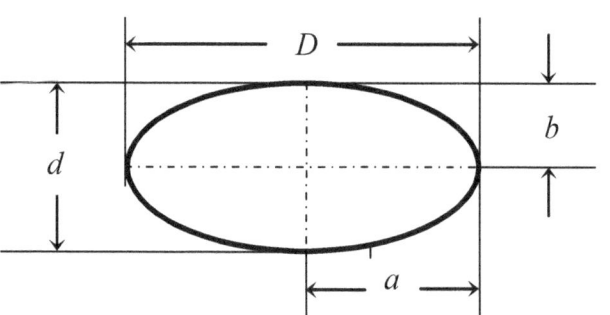

12.66) *Perimeter*: $P \approx \pi\sqrt{2(a^2 + b^2)}$

12.14 Segment of a Parabola

12.67) *Area*: $A = \dfrac{2}{3}ab$

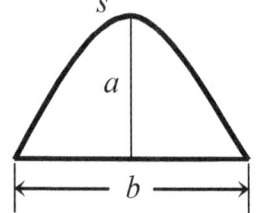

12.68) *Length of the arc*: $s = \dfrac{1}{2}\sqrt{b^2 + 16a^2} + \dfrac{b^2}{8a}\ln\left(\dfrac{4a + \sqrt{b^2 + 16a^2}}{b}\right)$

12.69) $P = s + b$

13.0 Solid Geometry

13.1 Cube

13.1) *Volume*: $V = a^3$

13.2) *Surface area*: $A_s = 6a^2$

13.3) $d = a\sqrt{3}$

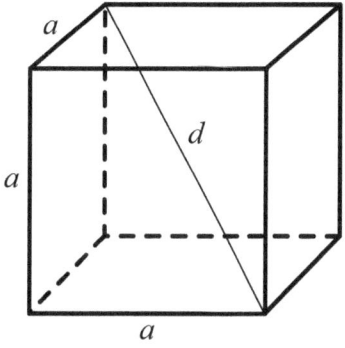

13.2 Rectangular Parallelepiped or Cuboid

13.4) *Volume*: $V = abc$

13.5) *Surface area*: $A_s = 2(ab + bc + ca)$

13.6) $d = \sqrt{a^2 + b^2 + c^2}$

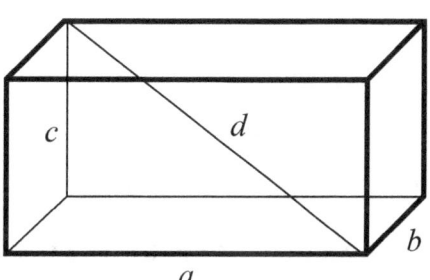

13.3 Parallelepiped
13.7) *Volume*: $V = Ah$
13.8) *Volume*: $V = abc\sin\theta$
13.9) *Surface area*:
  A_s = sum of areas of all the sides

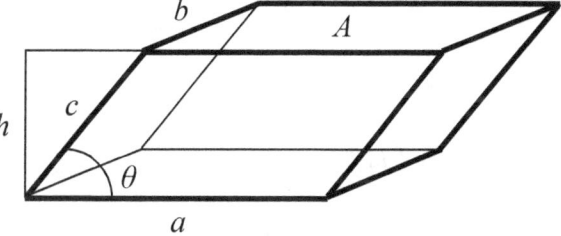

13.4 Cylinder, Solid
For right circular cylinder:
13.10) *Volume*: V = base area × length

13.11) *Volume*: $V = \pi r^2 l = \dfrac{\pi}{4} d^2 l$

13.12) *Lateral surface area*: A_l = circumference × length
13.13) *Lateral surface area*: $A_l = 2\pi r l = \pi d l$
13.14) *Total surface area*: $A_s = 2\pi r^2 + 2\pi r l = 2\pi r(r + l)$

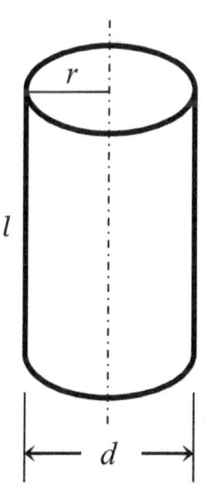

45

13.5 Cylinder, Hollow

13.15) *Volume:* $V = \dfrac{\pi}{4}(D^2 - d^2)l = \pi(R^2 - r^2)l$

13.16) *Lateral surface area (inside):* $A_l = \pi d l = 2\pi r l$
13.17) *Lateral surface area (outside):* $A_l = \pi D l = 2\pi R l$
13.18) *Total surface area:* $A_s = 2\pi r l + 2\pi R l + 2\pi(R^2 - r^2)$

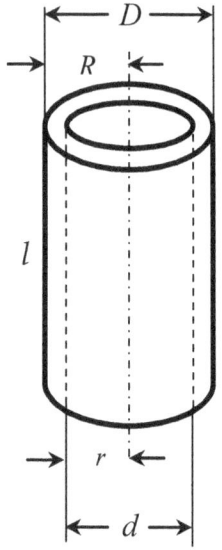

13.6 Sliced Cylinder or Cylinder with Oblique Face

13.19) *Volume:* $V = \dfrac{\pi r^2}{2}(l_1 + l_2) = \pi r^2 l$

13.20) *Lateral surface area:* $A_l = \pi r(l_1 + l_2) = 2\pi r l$

13.21) *Area:* $A = \pi r \sqrt{r^2 + \left(\dfrac{l_1 - l_2}{2}\right)^2}$

13.22) $l = \dfrac{l_1 + l_2}{2}$

13.23) *Total surface area:*

$A_s = \pi r(l_1 + l_2) + \pi r^2 + \pi r \sqrt{r^2 + \left(\dfrac{l_1 - l_2}{2}\right)^2}$

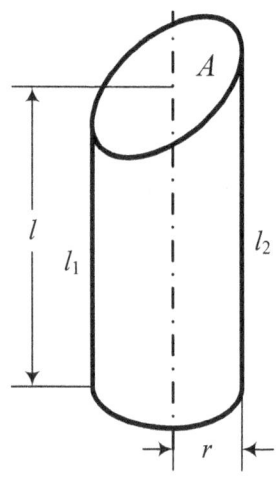

13.7 Ungula or Cylindrical Hoof

13.24) *Volume:* $V = \dfrac{2}{3} r^2 l$

13.25) *Lateral surface area:* $A_l = 2rl$

13.26) *Total surface area:* $A_s = 2rl + \dfrac{\pi r^2}{2} + \dfrac{\pi}{2} r \sqrt{r^2 + l^2}$

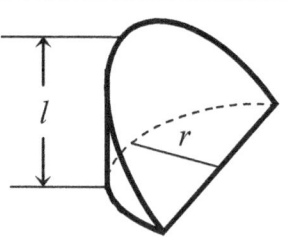

13.8 Cylinder with Slant Height

13.27) *Volume:* $V = \pi r^2 h$

13.28) *Volume:* $V = \pi r^2 l \sin\theta$

13.29) *Lateral surface area:* $A_l = 2\pi r l$

13.30) *Lateral surface area:* $A_l = 2\pi r h \dfrac{1}{\sin\theta}$

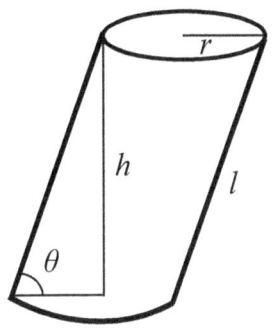

13.9 Cone

13.31) *Volume:* $V = \dfrac{1}{3} \times \text{base area} \times \text{height}$

13.32) *Volume:* $V = \dfrac{\pi}{3} r^2 h$

13.33) *Lateral surface area:* $A_l = \pi r l = \pi r \sqrt{r^2 + h^2}$

13.34) *Total surface area:* $A_s = \pi r^2 + \pi r l$

13.35) $l = \sqrt{r^2 + h^2}$

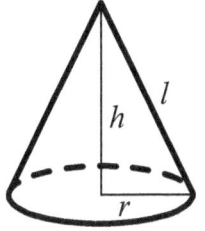

13.10 Frustum of a Cone

13.36) *Volume:* $V = \dfrac{\pi}{3} h(R^2 + rR + r^2) = \dfrac{\pi}{12} h(D^2 + Dd + d^2)$

13.37) *Lateral surface area:* $A_l = \pi(R+r)l = \dfrac{\pi}{2} l(D+d)$

13.38) *Lateral surface area:* $A_l = \pi(R+r)\sqrt{(R-r)^2 + h^2}$

13.39) *Total surface area:* $A_s = \pi R^2 + \pi(R+r)l + \pi r^2$

13.40) $l = \sqrt{(R-r)^2 + h^2} = \sqrt{\left(\dfrac{D-d}{2}\right)^2 + h^2}$

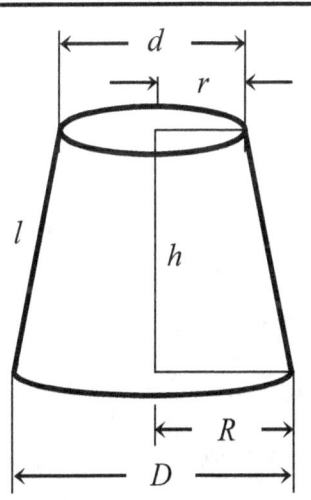

13.11 Prism

13.41) *Volume = base area × height*

13.42) *Lateral surface area = perimeter of the base × height*

13.11.1 Regular Triangular Prism

13.43) *Volume = area of the triangular region × h*
For area of a triangle, see Section 12.6.

13.44) *Total surface area =
sum of the surface areas of all the sides*

13.45) *Lateral surface area =
perimeter of the triangular region × h*

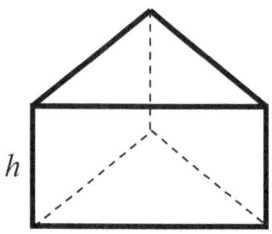

13.11.2 Square Prism

13.46) *Volume = area of the square region × h*

13.47) *Lateral surface area = perimeter of the square region × h*

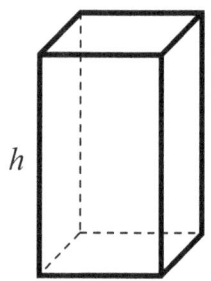

13.11.3 Prism with Polygon as Base

13.48) *Volume = area of the base polygon × h*

13.49) *Lateral surface = perimeter of the base polygon × h*

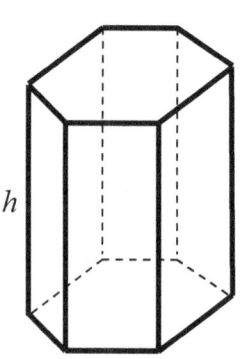

13.12 Pyramid

13.50) *Volume = $\frac{1}{3}$ × base area × height*

13.51) *Total surface area = sum of the areas of all the surfaces*

13.12.1 Triangular Pyramid

13.52) *Volume = $\frac{1}{3}$ × area of the triangular base × h*

13.53) *Total surface area = sum of the areas of all the four triangles*

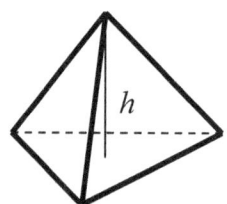

13.12.2 Square Pyramid

13.54) $Volume = \dfrac{1}{3} \times area\ of\ the\ square\ region \times h$

13.55) *Total surface area =*
 sum of the areas of the four triangles + area of the square

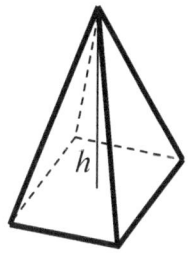

13.12.3 Pyramid with Polygon as Base

13.56) $Volume = \dfrac{1}{3} \times area\ of\ the\ base\ polygon \times h$

13.57) *Total surface area =*
 sum of the areas of all the triangles + area of base polygon

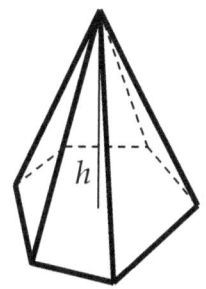

13.13 Frustum of a Pyramid or Cone

13.58) $Volume: V = \dfrac{1}{3} h \left(A_B + A_T + (A_B A_T)^{\frac{1}{2}} \right)$

13.59) *Lateral surface of the regular pyramid (or cone)*
$= \dfrac{1}{2} \times (perimeter\ of\ the\ base + perimeter\ of\ the\ top) \times l$

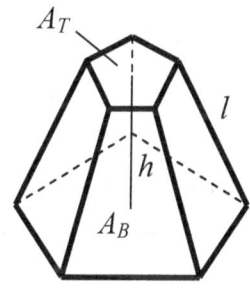

13.14 Barrel

13.60) $Volume: V = \dfrac{\pi}{12} h(2D^2 + d^2)$

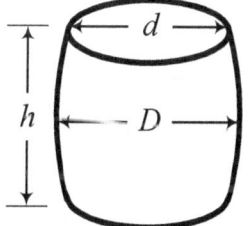

13.15 Sphere

13.61) $Volume: V = \dfrac{4}{3}\pi r^3$

13.62) $Volume: V = \dfrac{1}{6}\pi d^3$

13.63) $Surface\ area: A_s = 4\pi r^2 = \pi d^2$

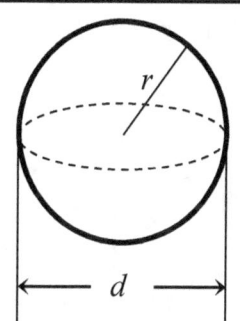

13.16 Zone of a Sphere

13.64) *Volume*: $V = \dfrac{\pi}{6} h(3r_1^2 + 3r_2^2 + h^2)$

13.65) *Lateral surface*: $A_l = 2\pi rh$

13.66) *Total surface area*: $A_s = \pi(2rh + r_1^2 + r_2^2)$

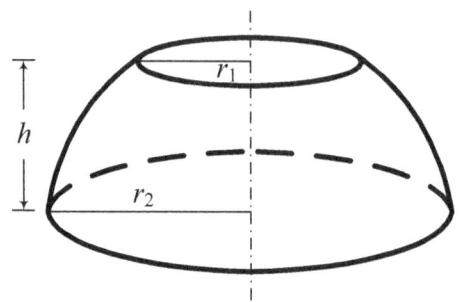

13.17 Segment of a Sphere or Spherical Cap

13.67) *Volume*: $V = \dfrac{\pi}{24} h(3b^2 + 4h^2) = \dfrac{1}{3}\pi h^2(3r - h)$

13.68) *Lateral surface*: $A_l = 2\pi rh = \dfrac{\pi}{4}(b^2 + 4h^2)$

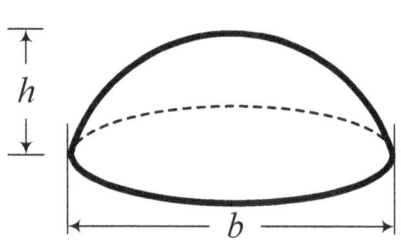

13.18 Sector of a Sphere

13.69) *Volume*: $V = \dfrac{2}{3}\pi r^2 h$

13.70) *Total surface area* $= A_s = \dfrac{\pi}{2} r(4h + b)$

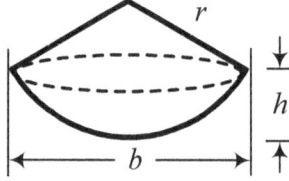

13.19 Sphere with Cylindrical Boring

13.71) *Volume*: $V = \dfrac{\pi}{6} h^3$

13.72) *Total surface area*: $A_s = 2\pi h(R + r)$

13.73) $h = 2\sqrt{R^2 - r^2}$

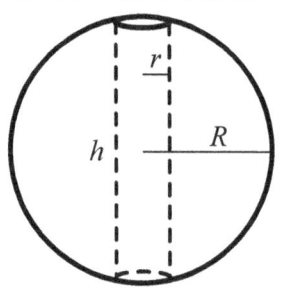

13.20 Spherical Triangle

13.74) *Area of the triangle*:

$A = (\alpha + \beta + \gamma - \pi)r^2$ (angles in radian)

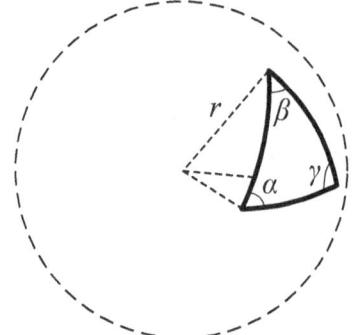

13.21 Torus or Anchor Ring

13.75) *Volume*: $V = 2\pi^2 R r^2$

13.76) *Volume*: $V = \dfrac{1}{4}\pi^2 (a+b)(b-a)^2$

13.77) *Total surface area*: $A_s = 4\pi^2 R r$

13.78) *Total surface area*: $A_s = \pi^2 (b^2 - a^2)$

13.79) $r = \dfrac{b-a}{2}$

13.80) $R = \dfrac{b+a}{2}$

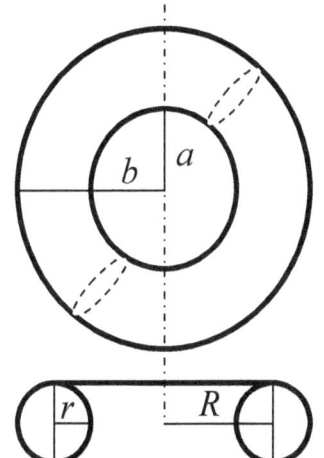

13.22 Ellipsoid

13.81) *Volume*: $V = \dfrac{4}{3}\pi abc$

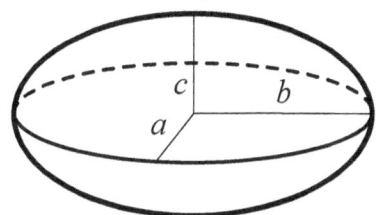

13.23 Paraboloid

13.82) *Volume*: $V = \dfrac{1}{2}\pi b^2 a$

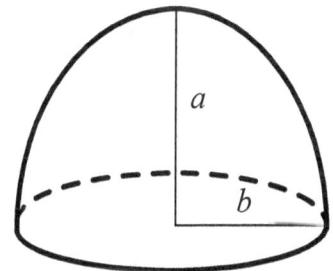

14.0 Analytical Geometry

14.1 Distance between Two Points on a Straight Line

14.1) $d = |P_1 P_2| = \sqrt{(x_2 - x_1)^2 + (y_2 - y_1)^2}$

14.2 Mid-Point of Two Points on a Straight Line

Coordinates of the mid-point of two points lying on a straight line are

14.2) $(x, y) = \left(\dfrac{x_1 + x_2}{2}, \dfrac{y_1 + y_2}{2} \right)$

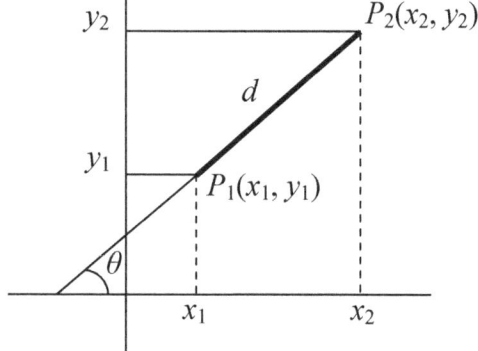

14.3 Slope of a Straight Line

14.3) $m = \dfrac{y_2 - y_1}{x_2 - x_1}$

14.4) $m = \tan \theta$

14.4 Equation of a Straight Line

14.4.1 Two-Point Form

14.5) $\dfrac{x - x_1}{x_2 - x_1} = \dfrac{y - y_1}{y_2 - y_1}$

14.4.2 Point-Slope Form

14.6) $y - y_1 = m(x - x_1)$

14.4.3 Slope-Intercept Form

14.7) $y = mx + c$

14.8) $c = \dfrac{x_2 y_1 - x_1 y_2}{x_2 - x_1}$

14.4.4 Intercepts Form

14.9) $\dfrac{x}{a} + \dfrac{y}{b} = 1$

14.4.5 Normal Form

14.10) $x\cos\theta + y\sin\theta = p$

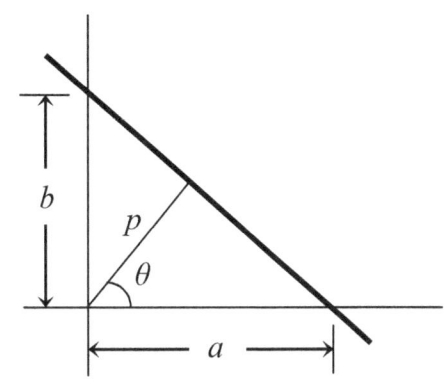

14.4.6 Line Parallel to the x-axis

14.11) $y = d$

where d is constant distance of the line from the x-axis.

14.4.7 Line Parallel to the y-axis

14.12) $x = d$

where d is constant distance of the line from the y-axis.

14.4.8 General Equation of a Straight Line

14.13) $ax + by + c = 0$

14.14) *Slope*: $m = -\dfrac{a}{b}$

14.15) *y-intercept* $= -\dfrac{c}{b}$

14.5 Ratio Formula

If point $P(x, y)$ divides the line joining the points $P_1(x_1, y_1)$ and $P_2(x_2, y_2)$ internally in the ratio $\dfrac{k_1}{k_2} = \dfrac{P_1P}{PP_2}$ then the coordinates of $P(x, y)$ are

14.16) $(x, y) = \left(\dfrac{k_1 x_2 + k_2 x_1}{k_1 + k_2}, \dfrac{k_1 y_2 + k_2 y_1}{k_1 + k_2} \right)$

If point $P(x, y)$ divides the line joining the points $P_1(x_1, y_1)$ and $P_2(x_2, y_2)$ externally in the ratio $\dfrac{k_1}{k_2} = \dfrac{P_1 P}{P_2 P}$ then the coordinates of $P(x, y)$ are

14.17) $(x, y) = \left(\dfrac{k_1 x_2 - k_2 x_1}{k_1 - k_2}, \dfrac{k_1 y_2 - k_2 y_1}{k_1 - k_2} \right)$

14.6 Condition of Parallelism and Perpendicularity of Two Lines

Two lines are parallel, when

14.18) $m_1 = m_2$ or $\alpha_1 = \alpha_2$

where m_i and α_i are slope of line and angle formed by line with abscissa, respectively.

Two lines are perpendicular, when

14.19) $m_1 m_2 = -1$

For two lines, $a_1 x + b_1 y + c_1 = 0$ and $a_2 x + b_2 y + c_2 = 0$, the lines will be parallel, if

14.20) $-\dfrac{a_1}{b_1} = -\dfrac{a_2}{b_2}$ or $\dfrac{a_1}{a_2} = \dfrac{b_1}{b_2}$, and the lines are perpendicular, if

14.21) $\left(-\dfrac{a_1}{b_1} \right)\left(-\dfrac{a_2}{b_2} \right) = -1$ or

14.22) $a_1 a_2 + b_1 b_2 = 0$

The two lines would intersect only, if

14.23) $a_1 b_2 - a_2 b_1 \neq 0$, and point of intersection (x, y) is

14.24) $(x, y) = \left(\dfrac{b_1 c_2 - b_2 c_1}{a_1 b_2 - a_2 b_1},\ \dfrac{c_1 a_2 - c_2 a_1}{a_1 b_2 - a_2 b_1} \right)$

Angle made by two intersecting lines is

14.25) $\tan \varphi = \dfrac{m_2 - m_1}{1 + m_2 m_1}$

14.26) $\tan \varphi = \tan(\theta_2 - \theta_1)$

14.27) $\tan \varphi = \dfrac{\tan \theta_2 - \tan \theta_1}{1 + \tan \theta_2 \tan \theta_1}$

14.28) $\varphi = \tan^{-1} \left| \dfrac{m_2 - m_1}{1 + m_2 m_1} \right|$

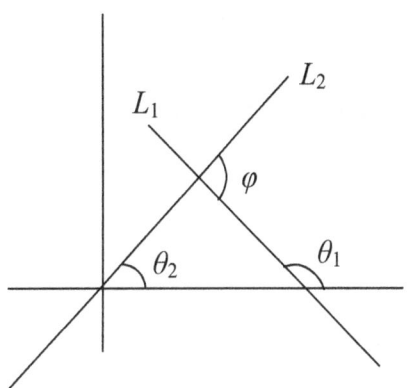

14.7 Condition of Concurrency of Three Lines

14.29) $\begin{array}{l} a_1 x + b_1 y + c_1 = 0 \\ a_2 x + b_2 y + c_2 = 0 \\ a_3 x + b_3 y + c_3 = 0 \end{array}$

The lines (Eq. 14.29) will be concurrent, if

14.30) $a_3(b_1 c_2 - b_2 c_1) + b_3(c_1 a_2 - c_2 a_1) + c_3(a_1 b_2 - a_2 b_1) = 0$

14.31) $\begin{vmatrix} a_1 & b_1 & c_1 \\ a_2 & b_2 & c_2 \\ a_3 & b_3 & c_3 \end{vmatrix} = 0$

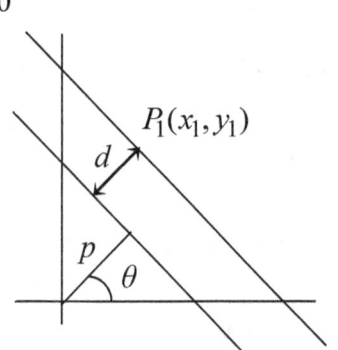

14.8 Distance of a Point from a Line

Distance of a point $P_1(x_1, y_1)$ from a given line

$x\cos\theta + y\sin\theta = p$ is

14.32) $d = |x_1 \cos\theta + y_1 \sin\theta - p|$

Distance of a point $P_1(x_1, y_1)$ from a given line $ax + bx + c = 0$ is

14.33) $d = \dfrac{|ax_1 + by_1 + c|}{\sqrt{a^2 + b^2}}$

14.9 Area of a Triangular Region

14.34) Area: $A = \dfrac{1}{2}[x_1(y_2 - y_3) + x_2(y_3 - y_1) + x_3(y_1 - y_2)]$

14.35) Area: $A = \dfrac{1}{2}\begin{vmatrix} x_1 & y_1 & 1 \\ x_2 & y_2 & 1 \\ x_3 & y_3 & 1 \end{vmatrix}$

14.10 Condition of Collinearity (on a Single Line) of Three Points

Three points $P_1(x_1, y_1)$, $P_2(x_2, y_2)$, and $P_3(x_3, y_3)$ are collinear, when

14.36) $\begin{vmatrix} x_1 & y_1 & 1 \\ x_2 & y_2 & 1 \\ x_3 & y_3 & 1 \end{vmatrix} = 0$

14.37) $x_1(y_2 - y_3) + x_2(y_3 - y_1) + x_3(y_1 - y_2) = 0$

14.11 Circle

Equation of a circle with given radius and center:

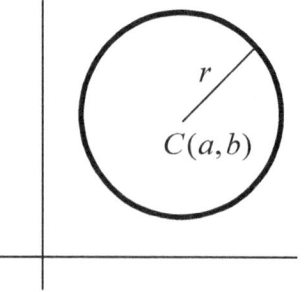

14.38) $(x - a)^2 + (y - b)^2 = r^2$

Equation of a circle with center at the origin and with given radius:

14.39) $x^2 + y^2 = r^2$

General equation of a circle is

14.40) $x^2 + y^2 + 2gx + 2fy + c = 0$ (center at $(-g, -f)$)

14.41) $r = \sqrt{g^2 + f^2 - c}$

Equation of a circle having points $A(x_1, y_1)$ and $B(x_2, y_2)$ as the extremities of its diameter is

14.42) $(x - x_1)(x - x_2) + (y - y_1)(y - y_2) = 0$

Tangent to a circle at point $P_1(x_1, y_1)$ is

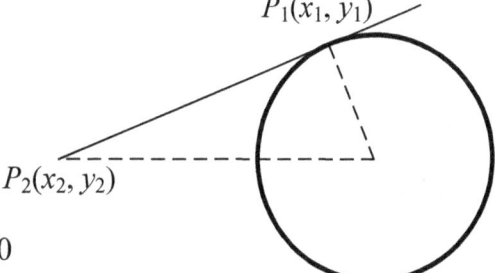

14.43) $xx_1 + yy_1 + g(x + x_1) + f(y + y_1) + c = 0$

Length of the tangent from point $P_2(x_2, y_2)$ to a circle ($x^2 + y^2 + 2gx + 2fy + c = 0$) is

14.44) $d = \sqrt{x_2^2 + y_2^2 + 2gx_2 + 2fy_2 + c}$

14.12 Ellipse

14.45) $\overline{PF'} + \overline{PF} = 2a$

14.46) $c = \sqrt{a^2 - b^2}$

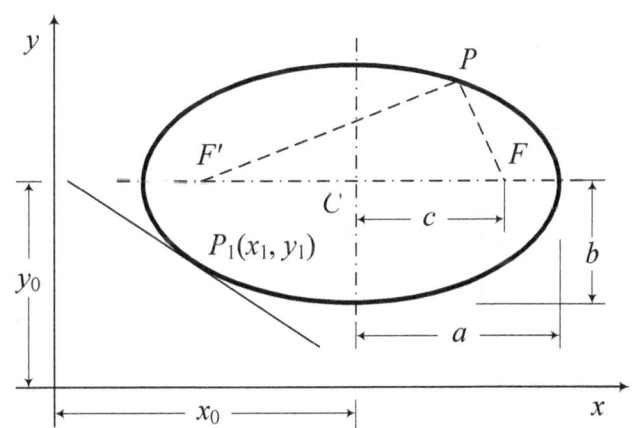

Equation of an ellipse with center $C(x_0, y_0)$ is

14.47) $\dfrac{(x - x_0)^2}{a^2} + \dfrac{(y - y_0)^2}{b^2} = 1$

Equation of an ellipse when center is at the origin is

14.48) $\dfrac{x^2}{a^2} + \dfrac{y^2}{b^2} = 1$

Tangent to an ellipse at a point $P_1(x_1, y_1)$ is

14.49) $y = -\dfrac{b^2}{a^2} \dfrac{(x_1 - x_0)(x - x_1)}{y_1 - y_0} + y_1$

14.13 Parabola

Equation of a parabola with center $C(x_0, y_0)$ is

14.50) If parabola opens to right: $(y - y_0)^2 = 4p(x - x_0)$

14.51) If parabola opens to left: $(y - y_0)^2 = -4p(x - x_0)$

14.52) If parabola opens upward: $(x - x_0)^2 = 4p(y - y_0)$

14.53) If parabola opens downward: $(x - x_0)^2 = -4p(y - y_0)$

Equation of a parabola when vertex is at the origin is

14.54) If parabola opens to right:
$y^2 = 4px$

14.55) If parabola opens to left:
$y^2 = -4px$

14.56) If parabola opens upward:
$x^2 = 4py$

14.57) If parabola opens downward:
$x^2 = -4py$

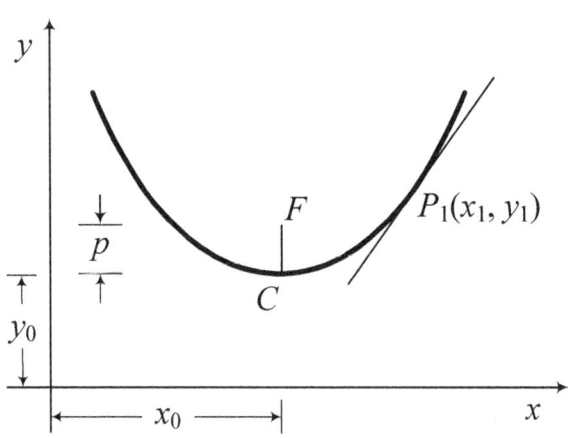

Tangent to a parabola at point $P_1(x_1, y_1)$ is

14.58) $y = \dfrac{2(y_1 - y_0)(x - x_1)}{(x_1 - x_0)} + y_1$

14.14 Hyperbola

Equation of hyperbola with center $C(x_0, y_0)$ is

14.59) $\dfrac{(x - x_0)^2}{a^2} - \dfrac{(y - y_0)^2}{b^2} = 1$

Equation of hyperbola at the origin is

14.60) $\dfrac{x^2}{a^2} - \dfrac{y^2}{b^2} = 1$

14.61) $c = \sqrt{a^2 + b^2}$

14.62) $\overline{F'P} - \overline{FP} = 2a$

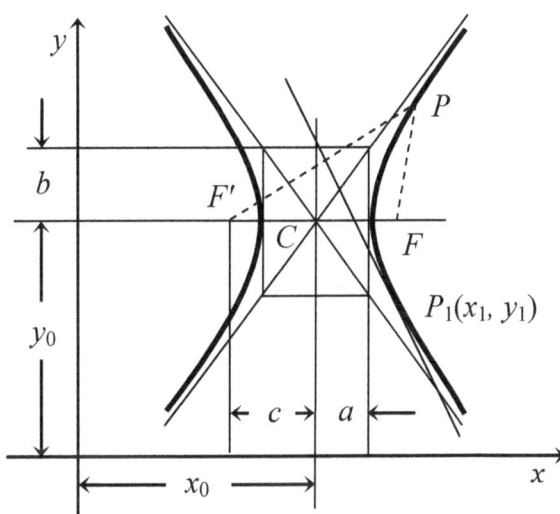

Slopes of the asymptotes are

14.63) $m = \pm \dfrac{b}{a}$

Tangent to hyperbola at point $P_1(x_1, y_1)$ is

14.64) $y = \dfrac{b^2}{a^2} \dfrac{(x_1 - x_0)(x - x_1)}{y_1 - y_0} + y_1$

15.0 Derivatives

15.1) $\dfrac{d}{dx}(a) = 0$

15.2) $\dfrac{d}{dx}(x) = 1$

15.3) $\dfrac{d}{dx}(ax) = a$

15.4) $\dfrac{d}{dx}\left(\dfrac{1}{x}\right) = -\dfrac{1}{x^2}$

15.5) $\dfrac{d}{dx}(\sqrt{x}) = \dfrac{1}{2\sqrt{x}}$

15.6) $\dfrac{d}{dx}(x^n) = nx^{n-1}$

15.7) $\dfrac{d}{dx}(ax^n) = anx^{n-1}$

15.8) $\dfrac{d}{dx}(x^x) = x^x(1 + \ln x)$

15.9) $\dfrac{d}{dx}(au) = a\dfrac{du}{dx}$ $u = f(x)$

15.10) $\dfrac{d}{dx}(u^n) = nu^{n-1}\dfrac{du}{dx}$ $u = f(x)$

15.11) $\dfrac{d}{dx}(u+v) = \dfrac{du}{dx} + \dfrac{dv}{dx}$ $u = f(x)$ and $v = f(x)$

15.12) $\dfrac{d}{dx}(uv) = u\dfrac{dv}{dx} + v\dfrac{du}{dx}$ $u = f(x)$ and $v = f(x)$

15.13) $\dfrac{d}{dx}(uvw) = uv\dfrac{dw}{dx} + vw\dfrac{du}{dx} + wu\dfrac{dv}{dx}$ $u = f(x)$, $v = f(x)$, and $w = f(x)$

15.14) $\dfrac{d}{dx}(u^v) = u^v\left(\dfrac{1}{u}\dfrac{du}{dx}v + \dfrac{dv}{dx}\ln u\right)$ $u = f(x)$ and $v = f(x)$

15.15) $\dfrac{d}{dx}\left(\dfrac{u}{v}\right) = \dfrac{v\dfrac{du}{dx} - u\dfrac{dv}{dx}}{v^2}$ $u = f(x)$ and $v = f(x)$

15.16) $\dfrac{d}{dx}e^x = e^x$

15.17) $\dfrac{d}{dx}e^{-x} = -e^{-x}$

15.18) $\dfrac{d}{dx}e^{ax} = ae^{ax}$

15.19) $\dfrac{d}{dx}\sqrt{e^x} = \dfrac{1}{2}\sqrt{e^x}$

15.20) $\dfrac{d}{dx}e^u = e^u\dfrac{du}{dx}$ $u = f(x)$

15.21) $\dfrac{d}{dx}(a^x) = a^x \ln a$

15.22) $\dfrac{d}{dx}(a^{nx}) = na^{nx}\ln a$

15.23) $\dfrac{d}{dx}(a^u) = a^u \ln a \dfrac{du}{dx}$ $u = f(x)$

15.24) $\dfrac{d}{dx}(\ln x) = \dfrac{1}{x}$

15.25) $\dfrac{d}{dx}(\ln u) = \dfrac{1}{u}\dfrac{du}{dx}$ $u = f(x)$

15.26) $\dfrac{d}{dx}\ln(1+x) = \dfrac{1}{1+x}$

15.27) $\dfrac{d}{dx}\ln(1-x) = \dfrac{-1}{1-x}$

15.28) $\dfrac{d}{dx}(\ln x^n) = \dfrac{n}{x}$

15.29) $\dfrac{d}{dx}\ln\sqrt{x} = \dfrac{1}{2x}$

15.30) $\dfrac{d}{dx}\ln u^n = \dfrac{n}{u}\dfrac{du}{dx}$ $u=f(x)$

15.31) $\dfrac{d}{dx}\log_a x = \dfrac{1}{x\ln a}$

15.32) $\dfrac{d}{dx}\log_a u = \dfrac{\log_a e}{u}\dfrac{du}{dx} = \dfrac{1}{u\ln a}\dfrac{du}{dx}$ $u=f(x)$

15.33) $\dfrac{d}{dx}\sin x = \cos x$

15.34) $\dfrac{d}{dx}\cos x = -\sin x$

15.35) $\dfrac{d}{dx}\tan x = \sec^2 x = 1+\tan^2 x$

15.36) $\dfrac{d}{dx}\sin^n x = n(\sin^{n-1} x)\cos x$

15.37) $\dfrac{d}{dx}\cos^n x = -n(\cos^{n-1} x)\sin x$

15.38) $\dfrac{d}{dx}\tan^n x = n(\tan^{n-1} x)(1+\tan^2 x)$

15.39) $\dfrac{d}{dx}\sinh x = \cosh x$

15.40) $\dfrac{d}{dx}\cosh x = \sinh x$

15.41) $\dfrac{d}{dx}\tanh x = \dfrac{1}{\cosh^2 x}$

15.42) $\dfrac{d}{dx}\sin^{-1} x = \dfrac{1}{\sqrt{1-x^2}}$

15.43) $\dfrac{d}{dx}\cos^{-1} x = -\dfrac{1}{\sqrt{1-x^2}}$

15.44) $\dfrac{d}{dx}\tan^{-1} x = \dfrac{1}{1+x^2}$

Second derivative can be represented as

15.45) $\dfrac{d}{dx}\left(\dfrac{dy}{dx}\right) = \dfrac{d^2 y}{dx^2} = f''(x) = y'' = f^{(2)}(x)$

15.46) $\dfrac{d^2(uv)}{dx^2} = u\dfrac{d^2 v}{dx^2} + 2\dfrac{du}{dx}\dfrac{dv}{dx} + v\dfrac{d^2 u}{dx^2}$

Third derivative can be represented as

15.47) $\dfrac{d}{dx}\left(\dfrac{d^2 y}{dx^2}\right) = \dfrac{d^3 y}{dx^3} = f'''(x) = y''' = f^{(3)}(x)$

61

15.48) $\dfrac{d^3(uv)}{dx^3} = u\dfrac{d^3v}{dx^3} + 3\dfrac{du}{dx}\dfrac{d^2v}{dx^2} + 3\dfrac{d^2u}{dx^2}\dfrac{dv}{dx} + v\dfrac{d^3u}{dx^3}$

Slope of a curve at $x = a$ is

15.49) $Slope = \dfrac{dy}{dx}\bigg|_{x=a}$ $\qquad y = f(x)$

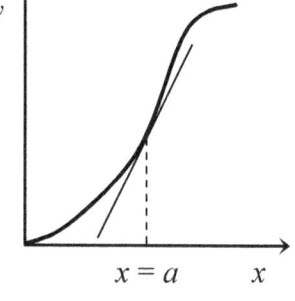

The differential of $f(x, y)$ is

15.50) $df = \dfrac{\partial f}{\partial x} dx + \dfrac{\partial f}{\partial y} dy$

15.51) $\dfrac{\partial^2 f}{\partial x \partial y} = \dfrac{\partial}{\partial x}\left(\dfrac{\partial f}{\partial y}\right)$ $\qquad$ 15.52) $\dfrac{\partial^2 f}{\partial x^2} = \dfrac{\partial}{\partial x}\left(\dfrac{\partial f}{\partial x}\right)$

15.53) $\left(\dfrac{\partial x}{\partial y}\right)_z \left(\dfrac{\partial y}{\partial z}\right)_x \left(\dfrac{\partial z}{\partial x}\right)_y = -1$ $\qquad$ (Cyclic rule)

15.54) $\left(\dfrac{\partial x}{\partial y}\right)_z = \dfrac{1}{\left(\dfrac{\partial y}{\partial x}\right)_z}$

16.0 Integrals

16.1) $\int 1 \cdot dx = x$

16.2) $\int a\, dx = ax$

16.3) $\int x\, dx = \dfrac{x^2}{2}$

16.4) $\int \sqrt{x}\, dx = \dfrac{2}{3}\sqrt{x^3}$

16.5) $\int \dfrac{1}{\sqrt{x}}\, dx = 2\sqrt{x}$

16.6) $\int x^n\, dx = \dfrac{x^{n+1}}{n+1}$

16.7) $\int (ax+b)^n\, dx = \dfrac{(ax+b)^{n+1}}{a(n+1)}$

16.8) $\int \dfrac{1}{x}\, dx = \ln x$

16.9) $\int (u+v)\, dx = \int u\, dx + \int v\, dx \qquad u = f(x) \text{ and } v = f(x)$

16.10) $\int \dfrac{(du/dx)}{u}\, dx = \ln u \qquad u = f(x)$

16.11) $\int u^n \dfrac{du}{dx}\, dx = \dfrac{u^{n+1}}{n+1} \qquad u = f(x)$

16.12) $\int u \dfrac{dv}{dx}\, dx = uv - \int v \dfrac{du}{dx}\, dx \qquad u = f(x) \text{ and } v = f(x)$

16.13) $\int a^x\, dx = \dfrac{a^x}{\ln a}$

16.14) $\int a^{bx}\, dx = \dfrac{a^{bx}}{b \ln a}$

16.15) $\int e^x\, dx = e^x$

16.16) $\int e^{ax}\, dx = \dfrac{e^{ax}}{a}$

16.17) $\int x e^{ax}\, dx = \dfrac{e^{ax}}{a}\left(x - \dfrac{1}{a}\right)$

16.18) $\int \ln x\, dx = x \ln x - x$

16.19) $\int x \ln x\, dx = \dfrac{x^2}{2}\left(\ln x - \dfrac{1}{2}\right)$

16.20) $\int \dfrac{\ln x}{x}\, dx = \dfrac{1}{2} \ln^2 x$

16.21) $\int \dfrac{1}{x \ln x}\, dx = \ln(\ln x)$

16.22) $\int \dfrac{dx}{x-a} = \ln(x-a)$

16.23) $\int \dfrac{x\, dx}{x-a} = x + a \ln(x-a)$

16.24) $\int \dfrac{dx}{x+a} = \ln(x+a)$

16.25) $\int \dfrac{x\, dx}{x+a} = x - a \ln(x+a)$

16.26) $\int \dfrac{dx}{ax+b} = \dfrac{1}{a} \ln(ax+b)$

16.27) $\int \dfrac{x\, dx}{ax+b} = \dfrac{x}{a} - \dfrac{b}{a^2} \ln(ax+b)$

16.28) $\int \dfrac{dx}{x(ax+b)} = \dfrac{1}{b} \ln\left(\dfrac{x}{ax+b}\right)$

16.29) $\int \dfrac{dx}{x^2+a^2} = \dfrac{1}{a} \tan^{-1} \dfrac{x}{a}$

16.30) $\int \dfrac{x\, dx}{x^2+a^2} = \dfrac{1}{2} \ln(x^2+a^2)$

16.31) $\int \dfrac{dx}{x^2-a^2} = \dfrac{1}{2a}\ln\left(\dfrac{x-a}{x+a}\right)$

16.32) $\int \dfrac{xdx}{x^2-a^2} = \dfrac{1}{2}\ln(x^2-a^2)$

16.33) $\int \dfrac{dx}{a^2-x^2} = \dfrac{1}{2a}\ln\left(\dfrac{a+x}{a-x}\right)$

16.34) $\int \dfrac{xdx}{a^2-x^2} = -\dfrac{1}{2}\ln(a^2-x^2)$

16.35) $\int \dfrac{dx}{(x-a)(x-b)} = \dfrac{1}{a-b}\ln\dfrac{x-a}{x-b}$

16.36) $\int \dfrac{dx}{(ax+b)^2} = \dfrac{-1}{a(ax+b)}$

16.37) $\int \dfrac{xdx}{(ax+b)^2} = \dfrac{b}{a^2(ax+b)} + \dfrac{1}{a^2}\ln(ax+b)$

16.38) $\int \dfrac{dx}{(ax+b)(px+q)} = \dfrac{1}{bp-aq}\ln\left(\dfrac{px+q}{ax+b}\right)$

16.39) $\int \dfrac{xdx}{(ax+b)(px+q)} = \dfrac{1}{bp-aq}\left(\dfrac{b}{a}\ln(ax+b) - \dfrac{q}{p}\ln(px+q)\right)$

16.40) $\int \dfrac{dx}{\sqrt{ax+b}} = \dfrac{2\sqrt{ax+b}}{a}$

16.41) $\int \dfrac{xdx}{\sqrt{ax+b}} = \dfrac{2(ax-2b)\sqrt{ax+b}}{3a^2}$

16.42) $\int \dfrac{dx}{\sqrt{x^2+a^2}} = \ln(x+\sqrt{x^2+a^2})$

16.43) $\int \dfrac{xdx}{\sqrt{x^2+a^2}} = \sqrt{x^2+a^2}$

16.44) $\int \dfrac{dx}{\sqrt{x^2-a^2}} = \ln(x+\sqrt{x^2-a^2})$

16.45) $\int \dfrac{xdx}{\sqrt{x^2-a^2}} = \sqrt{x^2-a^2}$

16.46) $\int \dfrac{dx}{\sqrt{a^2-x^2}} = \sin^{-1}\dfrac{x}{a}$

16.47) $\int \dfrac{xdx}{\sqrt{a^2-x^2}} = -\sqrt{a^2-x^2}$

16.48) $\int \sqrt{x^2-a^2}\,dx = \dfrac{x\sqrt{x^2-a^2}}{2} - \dfrac{a^2}{2}\ln(x+\sqrt{x^2-a^2})$

16.49) $\int \sqrt{a^2-x^2}\,dx = \dfrac{x\sqrt{a^2-x^2}}{2} + \dfrac{a^2}{2}\sin^{-1}\dfrac{x}{a}$

16.50) $\int \sqrt{x^2+a^2}\,dx = \dfrac{x\sqrt{x^2+a^2}}{2} + \dfrac{a^2}{2}\ln(x+\sqrt{x^2+a^2})$

16.51) $\int \sin x\,dx = -\cos x$

16.52) $\int \cos x\,dx = \sin x$

16.53) $\int \tan x\, dx = -\ln\cos x$

16.54) $\int \sin ax\, dx = \dfrac{-\cos ax}{a}$

16.55) $\int \cos ax\, dx = \dfrac{\sin ax}{a}$

16.56) $\int \tan ax\, dx = -\dfrac{\ln\cos ax}{a}$

16.57) $\int \sinh ax\, dx = \dfrac{\cosh ax}{a}$

16.58) $\int \cosh ax\, dx = \dfrac{\sinh ax}{a}$

16.59) $\int \tanh ax\, dx = \dfrac{\ln\cosh ax}{a}$

16.60) $\int \sin^{-1} x\, dx = x\sin^{-1} x + \sqrt{1-x^2}$

16.61) $\int \cos^{-1} x\, dx = x\cos^{-1} x - \sqrt{1-x^2}$

16.62) $\int \tan^{-1} x\, dx = x\tan^{-1} x - \dfrac{1}{2}\ln(1+x^2)$

16.1 Area under the Curve

16.63) $Area = \displaystyle\int_a^b f(x)\, dx$

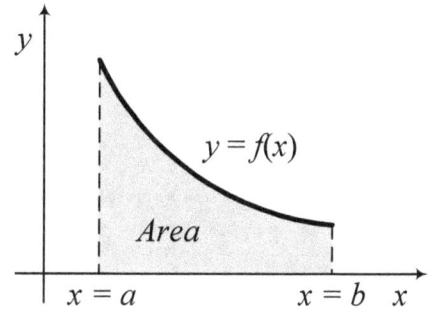

17.0 Vectors

17.1) $\vec{A} = |\vec{A}|\hat{e} = A\hat{e}$

where $\vec{A}$ is a vector, $|\vec{A}|$ or A is magnitude of the vector, and $\hat{e}$ is unit vector or direction of the vector.

Unit vector has magnitude of 1 and it is equal to

17.2) $\hat{e} = \dfrac{\vec{A}}{|\vec{A}|} = \dfrac{\vec{A}}{A}$

Sum of two vectors $\vec{A} = a_1\hat{i} + a_2\hat{j} + a_3\hat{k}$ and $\vec{B} = b_1\hat{i} + b_2\hat{j} + b_3\hat{k}$ is

17.3) $\vec{A} + \vec{B} = (a_1 + b_1)\hat{i} + (a_2 + b_2)\hat{j} + (a_3 + b_3)\hat{k}$

Difference of two vectors $\vec{A} = a_1\hat{i} + a_2\hat{j} + a_3\hat{k}$ and $\vec{B} = b_1\hat{i} + b_2\hat{j} + b_3\hat{k}$ is

17.4) $\vec{A} - \vec{B} = (a_1 - b_1)\hat{i} + (a_2 - b_2)\hat{j} + (a_3 - b_3)\hat{k}$

Multiplication of $\vec{A} = a_1\hat{i} + a_2\hat{j} + a_3\hat{k}$ by a scalar λ gives

17.5) $\lambda \vec{A} = \lambda a_1\hat{i} + \lambda a_2\hat{j} + \lambda a_3\hat{k}$

The magnitude or modulus of vector $\vec{A} = a_1\hat{i} + a_2\hat{j} + a_3\hat{k}$ is

17.6) $|\vec{A}| = A = \sqrt{a_1^2 + a_2^2 + a_3^2}$

17.1 Dot Product of Vectors

Dot product of two vectors $\vec{A} = a_1\hat{i} + a_2\hat{j} + a_3\hat{k}$ and $\vec{B} = b_1\hat{i} + b_2\hat{j} + b_3\hat{k}$ is

17.7) $\vec{A} \cdot \vec{B} = a_1b_1 + a_2b_2 + a_3b_3$

17.8) $\vec{A} \cdot \vec{B} = |\vec{A}||\vec{B}|\cos\theta = AB\cos\theta$

17.9) $\vec{A} \cdot \vec{B} = \vec{B} \cdot \vec{A}$

17.10) $\vec{A} \cdot (\vec{B} + \vec{C}) = \vec{A} \cdot \vec{B} + \vec{A} \cdot \vec{C}$

Angle between two vectors $\vec{A} = a_1\hat{i} + a_2\hat{j} + a_3\hat{k}$ and $\vec{B} = b_1\hat{i} + b_2\hat{j} + b_3\hat{k}$ is

17.11) $\theta = \cos^{-1}\dfrac{\vec{A} \cdot \vec{B}}{AB}$

If the angle is right angle, the two vectors are orthogonal.

Dot products of unit vectors are

17.12) $\hat{i} \cdot \hat{i} = 1$

17.13) $\hat{i} \cdot \hat{j} = 0$

17.14) $\hat{i} \cdot \hat{k} = 0$

17.15) $\hat{j} \cdot \hat{i} = 0$

17.16) $\hat{j} \cdot \hat{j} = 1$

17.17) $\hat{j} \cdot \hat{k} = 0$

17.18) $\hat{k} \cdot \hat{i} = 0$

17.19) $\hat{k} \cdot \hat{j} = 0$

17.20) $\hat{k} \cdot \hat{k} - 1$

17.2 Cross Product of Vectors

Cross product of vectors $\vec{A} = a_1\hat{i} + a_2\hat{j} + a_3\hat{k}$ and $\vec{B} = b_1\hat{i} + b_2\hat{j} + b_3\hat{k}$ is

17.21) $\vec{A} \times \vec{B} = |\vec{A}||\vec{B}|\sin\theta\,\hat{n} = AB\sin\theta\,\hat{n}$ or

17.22) $\vec{A} \times \vec{B} =$ determinant of matrix $\begin{pmatrix} \hat{i} & \hat{j} & \hat{k} \\ a_1 & a_2 & a_3 \\ b_1 & b_2 & b_3 \end{pmatrix} = \begin{vmatrix} \hat{i} & \hat{j} & \hat{k} \\ a_1 & a_2 & a_3 \\ b_1 & b_2 & b_3 \end{vmatrix}$

17.23) $\vec{A} \times \vec{B} = (a_2 b_3 - a_3 b_2)\hat{i} - (a_1 b_3 - a_3 b_1)\hat{j} + (a_1 b_2 - a_2 b_1)\hat{k}$

17.24) $\vec{A} \times \vec{B} = -\vec{B} \times \vec{A}$

Cross products of unit vectors are

17.25) $\hat{i} \times \hat{i} = 0$ 17.26) $\hat{i} \times \hat{j} = \hat{k}$

17.27) $\hat{i} \times \hat{k} = -\hat{j}$ 17.28) $\hat{j} \times \hat{i} = -\hat{k}$

17.29) $\hat{j} \times \hat{j} = 0$ 17.30) $\hat{j} \times \hat{k} = \hat{i}$

17.31) $\hat{k} \times \hat{i} = \hat{j}$ 17.32) $\hat{k} \times \hat{j} = -\hat{i}$

17.33) $\hat{k} \times \hat{k} = 0$

17.3 Scalar Triple Product of Vectors

Scalar triple product of three vectors such as $\vec{A} = a_1 \hat{i} + a_2 \hat{j} + a_3 \hat{k}$, $\vec{B} = b_1 \hat{i} + b_2 \hat{j} + b_3 \hat{k}$, and $\vec{C} = c_1 \hat{i} + c_2 \hat{j} + c_3 \hat{k}$ is

17.34) $\vec{A} \cdot (\vec{B} \times \vec{C}) = \begin{vmatrix} a_1 & a_2 & a_3 \\ b_1 & b_2 & b_3 \\ c_1 & c_2 & c_3 \end{vmatrix} = a_1(b_2 c_3 - c_2 b_3) - a_2(b_1 c_3 - c_1 b_3) + a_3(b_1 c_2 - c_1 b_2)$

17.35) $\vec{A} \cdot (\vec{B} \times \vec{C}) = \vec{B} \cdot (\vec{C} \times \vec{A}) = \vec{C} \cdot (\vec{A} \times \vec{B})$

17.4 Vector Triple Product of Vectors

Vector triple product of three vectors is

17.36) $(\vec{A} \times \vec{B}) \times \vec{C} = -\vec{C} \times (\vec{A} \times \vec{B})$

18.0 Coordinate Systems

18.1 Rectangular Coordinates or Cartesian Coordinates
18.1) Coordinates: (x, y, z)
18.2) $x = x$, $y = y$, $z = z$

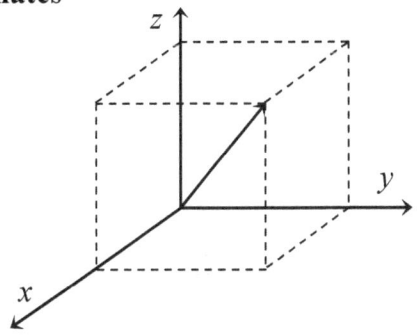

18.2 Cylindrical Coordinates
18.3) Coordinates: (r, θ, z)
18.4) $x = r\cos\theta$, $y = r\sin\theta$, $z = z$
18.5) $r = \sqrt{x^2 + y^2}$
18.6) $\theta = \tan^{-1}\dfrac{y}{x}$

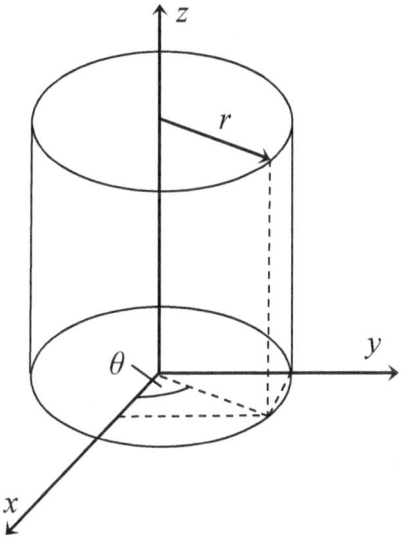

18.3 Spherical Coordinates
18.7) Coordinates: (r, θ, ϕ)
18.8) $x = r\sin\theta\cos\phi$, $y = r\sin\theta\sin\phi$, $z = r\cos\theta$
18.9) $r = \sqrt{x^2 + y^2 + z^2}$
18.10) $\theta = \cos^{-1}\left(\dfrac{z}{\sqrt{x^2 + y^2 + z^2}}\right)$
18.11) $\phi = \tan^{-1}\dfrac{y}{x}$

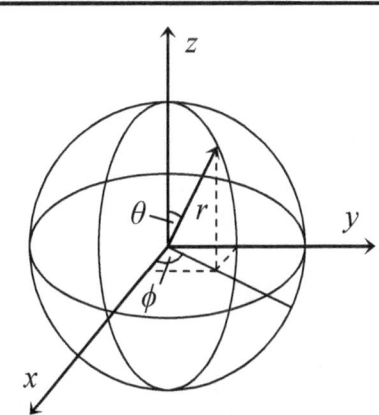

19.0 Gradient, Divergence, Curl, and Laplacian

19.1) The gradient of a scalar function T = grad $T = \nabla T$ is a vector quantity.
19.2) The divergence of a vector v = div $v = \nabla \cdot v$ is a scalar quantity.
19.3) The curl of a vector v = curl $v = \nabla \times v$ is a vector quantity.
19.4) The Laplacian of a scalar T = Laplacian $T = \nabla^2 T$ is a scalar quantity.

19.1 Gradient, Divergence, Curl, and Laplacian in Different Coordinate Systems

Cartesian or Rectangular	Cylindrical	Spherical
Coordinates Conversion		
$x = x$	$x = r\cos\theta$	$x = r\sin\theta\cos\phi$
$y = y$	$y = r\sin\theta$	$y = r\sin\theta\sin\phi$
$z = z$	$z = z$	$z = r\cos\theta$
Gradient of T, ∇T		
$\dfrac{\partial T}{\partial x}\hat{e}_x + \dfrac{\partial T}{\partial y}\hat{e}_y + \dfrac{\partial T}{\partial z}\hat{e}_z$	$\dfrac{\partial T}{\partial r}\hat{e}_r + \dfrac{1}{r}\dfrac{\partial T}{\partial \theta}\hat{e}_\theta + \dfrac{\partial T}{\partial z}\hat{e}_z$	$\dfrac{\partial T}{\partial r}\hat{e}_r + \dfrac{1}{r}\dfrac{\partial T}{\partial \theta}\hat{e}_\theta + \dfrac{1}{r\sin\theta}\dfrac{\partial T}{\partial \phi}\hat{e}_\phi$
Divergence of v, $\nabla \cdot v$		
$\dfrac{\partial v_x}{\partial x} + \dfrac{\partial v_y}{\partial y} + \dfrac{\partial v_z}{\partial z}$	$\dfrac{1}{r}\dfrac{\partial(rv_r)}{\partial r} + \dfrac{1}{r}\dfrac{\partial v_\theta}{\partial \theta} + \dfrac{\partial v_z}{\partial z}$	$\dfrac{1}{r^2}\dfrac{\partial(r^2 v_r)}{\partial r} + \dfrac{1}{r\sin\theta}\dfrac{\partial(v_\theta \sin\theta)}{\partial \theta}$
		$+ \dfrac{1}{r\sin\theta}\dfrac{\partial v_\phi}{\partial \phi}$
Curl of v, $\nabla \times v$		
$\left(\dfrac{\partial v_z}{\partial y} - \dfrac{\partial v_y}{\partial z}\right)\hat{e}_x + \left(\dfrac{\partial v_x}{\partial z} - \dfrac{\partial v_z}{\partial x}\right)\hat{e}_y$	$\left(\dfrac{1}{r}\dfrac{\partial v_z}{\partial \theta} - \dfrac{\partial v_\theta}{\partial z}\right)\hat{e}_r + \left(\dfrac{\partial v_r}{\partial z} - \dfrac{\partial v_z}{\partial r}\right)\hat{e}_\theta$	$\dfrac{1}{r\sin\theta}\left(\dfrac{\partial}{\partial \theta}(v_\phi \sin\theta) - \dfrac{\partial v_\theta}{\partial \phi}\right)\hat{e}_r$
$+ \left(\dfrac{\partial v_y}{\partial x} - \dfrac{\partial v_x}{\partial y}\right)\hat{e}_z$	$+ \dfrac{1}{r}\left(\dfrac{\partial(rv_\theta)}{\partial r} - \dfrac{\partial v_r}{\partial \theta}\right)\hat{e}_z$	$+ \left(\dfrac{1}{r\sin\theta}\dfrac{\partial v_r}{\partial \phi} - \dfrac{1}{r}\dfrac{\partial(rv_\phi)}{\partial r}\right)\hat{e}_\theta$
		$+ \dfrac{1}{r}\left(\dfrac{\partial(rv_\theta)}{\partial r} - \dfrac{\partial v_r}{\partial \theta}\right)\hat{e}_\phi$
Laplacian of T, $\nabla^2 T$		
$\dfrac{\partial^2 T}{\partial x^2} + \dfrac{\partial^2 T}{\partial y^2} + \dfrac{\partial^2 T}{\partial z^2}$	$\dfrac{1}{r}\dfrac{\partial}{\partial r}\left(r\dfrac{\partial T}{\partial r}\right)$	$\dfrac{1}{r^2}\dfrac{\partial}{\partial r}\left(r^2 \dfrac{\partial T}{\partial r}\right)$
	$+ \dfrac{1}{r^2}\dfrac{\partial^2 T}{\partial \theta^2} + \dfrac{\partial^2 T}{\partial z^2}$	$+ \dfrac{1}{r^2 \sin\theta}\dfrac{\partial}{\partial \theta}\left(\sin\theta \dfrac{\partial T}{\partial \theta}\right)$
		$+ \dfrac{1}{r^2 \sin^2\theta}\dfrac{\partial^2 T}{\partial \phi^2}$

20.0 Laplace Transform

The Laplace transform of a function $f(t)$ is $f(s)$ and $f(s) = \int_0^\infty f(t)e^{-st}dt$ or $f(s) = L\{f(t)\}$.

20.1 Table of Laplace Transforms

$f(s)$	$f(t)$	$f(s)$	$f(t)$
1	$\delta(t)$ (unit impulse)	$\dfrac{1}{(s+a)(s+b)}$	$-\dfrac{1}{a-b}\left(e^{-at}-e^{-bt}\right)$
$\dfrac{1}{s}$	$u(t)$ (unit step)	$\dfrac{1}{(s-a)(s-b)}$	$\dfrac{1}{a-b}\left(e^{at}-e^{bt}\right)$
$\dfrac{1}{s^2}$	t	$\dfrac{s}{(s+a)(s+b)}$	$\dfrac{1}{a-b}\left(ae^{-at}-be^{-bt}\right)$
$\dfrac{1}{s^n}$	$\dfrac{t^{n-1}}{(n-1)!}$	$\dfrac{s}{(s-a)(s-b)}$	$\dfrac{1}{a-b}\left(ae^{at}-be^{bt}\right)$
$\dfrac{1}{\sqrt{s}}$	$\dfrac{1}{\sqrt{\pi t}}$	$\dfrac{1}{s^2+a^2}$	$\dfrac{1}{a}\sin at$
e^{-as}	$\delta(t-a)$	$\dfrac{1}{s^2-a^2}$	$\dfrac{1}{a}\sinh at$
$\dfrac{1}{s+a}$	e^{-at}	$\dfrac{s}{s^2+a^2}$	$\cos at$
$\dfrac{1}{s-a}$	e^{at}	$\dfrac{s}{s^2-a^2}$	$\cosh at$
$\dfrac{1}{(s+a)^2}$	te^{-at}	$\dfrac{1}{s(s^2+a^2)}$	$\dfrac{1}{a^2}(1-\cos at)$
$\dfrac{1}{(s-a)^2}$	te^{at}	$\dfrac{1}{s(s^2-a^2)}$	$\dfrac{\cosh(at)-1}{a^2}$
$\dfrac{1}{(s+a)^n}$	$\dfrac{e^{-at}t^{n-1}}{(n-1)!}$	$\dfrac{1}{(s^2+a^2)^2}$	$\dfrac{\sin(at)-at\cos(at)}{2a^3}$
$\dfrac{1}{(s-a)^n}$	$\dfrac{e^{at}t^{n-1}}{(n-1)!}$	$\dfrac{1}{(s^2-a^2)^2}$	$\dfrac{at\cosh at-\sinh at}{2a^3}$
$\dfrac{1}{s(s+a)}$	$\dfrac{1-e^{-at}}{a}$	$\dfrac{s}{(s^2+a^2)^2}$	$\dfrac{t}{2a}(\sin at)$
$\dfrac{1}{s(s-a)}$	$\dfrac{e^{at}-1}{a}$	$\dfrac{s}{(s^2-a^2)^2}$	$\dfrac{t}{2a}(\sinh at)$

21.0 Statistics

21.1 Averages

21.1.1 Arithmetic Mean or Mean

The mean (arithmetic mean) of numbers such as $X_1, X_2, X_3, \ldots, X_N$ is

21.1) $\bar{X} = \dfrac{X_1 + X_2 + X_3 + \cdots + X_N}{N} = \dfrac{\sum_{i=1}^{i=N} X_i}{N}$

If numbers $X_1, X_2, X_3, \ldots, X_N$ occur $f_1, f_2, f_3, \ldots, f_N$ times respectively, then their mean (arithmetic mean) is

21.2) $\bar{X} = \dfrac{f_1 X_1 + f_2 X_2 + f_3 X_3 + \cdots + f_N X_N}{f_1 + f_2 + f_3 + \cdots + f_N} = \dfrac{\sum_{i=1}^{i=N} f_i X_i}{\sum_{i=1}^{i=N} f_i}$

21.3) $N = f_1 + f_2 + f_3 + \cdots + f_N = \sum_{i=1}^{i=N} f_i$ is the total number of observations.

21.1.1.1 Weighted Arithmetic Mean

If $w_1, w_2, w_3, \ldots, w_N$ are the weighting factors associated with $X_1, X_2, X_3, \ldots, X_N$, then arithmetic mean is

21.4) $\bar{X} = \dfrac{w_1 X_1 + w_2 X_2 + w_3 X_3 + \cdots + w_N X_N}{w_1 + w_2 + w_3 + \cdots + w_N} = \dfrac{\sum_{i=1}^{i=N} w_i X_i}{\sum_{i=1}^{i=N} w_i}$

21.1.2 Geometric Mean

The geometric mean of numbers such as $X_1, X_2, X_3, \ldots, X_N$ is

21.5) $$G = \sqrt[N]{X_1 \times X_2 \times X_3 \times \cdots \times X_N} = \left(\prod_{i=1}^{i=N} X_i\right)^{\frac{1}{N}}$$

If numbers $X_1, X_2, X_3, \ldots, X_N$ occur $f_1, f_2, f_3, \ldots, f_N$ times respectively, then their geometric mean is

21.6) $G = \sqrt[N]{X_1^{f_1} \times X_2^{f_2} \times X_3^{f_3} \times \cdots \times X_N^{f_N}}$ See Eq. 21.3

21.7) $\log G = \dfrac{1}{N}(f_1 \log X_1 + f_2 \log X_2 + f_3 \log X_3 + \cdots + f_N \log X_N) = \dfrac{1}{N}\sum_{i=1}^{i=N} f_i \log X_i$

21.1.3 Harmonic Mean

The harmonic mean of numbers such as $X_1, X_2, X_3, \ldots, X_N$ is

21.8) $$H = \dfrac{N}{\dfrac{1}{X_1} + \dfrac{1}{X_2} + \dfrac{1}{X_3} + \cdots + \dfrac{1}{X_N}} = \dfrac{1}{\dfrac{1}{N}\sum_{i=1}^{i=N}\dfrac{1}{X_i}}$$

21.9) $\dfrac{1}{H} = \dfrac{1}{N}\sum_{i=1}^{i=N}\dfrac{1}{X_i}$

If numbers $X_1, X_2, X_3, \ldots, X_N$ occur $f_1, f_2, f_3, \ldots, f_N$ times, respectively, then their harmonic mean is

21.10) $\dfrac{1}{H} = \dfrac{1}{N}\left(\dfrac{f_1}{X_1} + \dfrac{f_2}{X_2} + \dfrac{f_3}{X_3} + \cdots + \dfrac{f_N}{X_N}\right) = \dfrac{1}{N}\sum_{i=1}^{i=N}\dfrac{f_i}{X_i}$ See Eq. 21.3

21.1.4 Relationship between Arithmetic, Geometric, and Harmonic Means

21.11) $H \leq G \leq \overline{X}$

Equality sign exists only when all numbers $X_1, X_2, X_3, \ldots, X_N$ are identical.

21.1.5 Root Mean Square or Quadratic Mean

The root mean square or quadratic mean of numbers such as $X_1, X_2, X_3, \ldots, X_N$ is

$$21.12) \quad RMS = \sqrt{\frac{X_1^2 + X_2^2 + X_3^2 + \cdots + X_N^2}{N}} = \sqrt{\frac{\sum_{i=1}^{i=N} X_i^2}{N}}$$

If numbers $X_1, X_2, X_3, \ldots, X_N$ occur $f_1, f_2, f_3, \ldots, f_N$ times, respectively, then their root mean square is

$$21.13) \quad RMS = \sqrt{\frac{f_1 X_1^2 + f_2 X_2^2 + f_3 X_3^2 + \cdots + f_N X_N^2}{N}} = \sqrt{\frac{\sum_{i=1}^{i=N} f_i X_i^2}{N}} \quad \text{See Eq. 21.3}$$

21.1.6 Logarithmic Mean or Log Mean

If t_1, t_2 and T_1, T_2 are two sets of quantities, then logarithmic mean is

$$21.14) \quad \Delta_{lm} = \frac{(T_1 - t_1) - (T_2 - t_2)}{\ln\left(\dfrac{T_1 - t_1}{T_2 - t_2}\right)}$$

21.2 Median

21.15) N numbers such as $X_1, X_2, X_3, \ldots, X_N$ when arranged in the order of their magnitude, the middle value or arithmetic mean of two middle values is the median of N numbers.

Median of grouped data (data arranged in a frequency distribution) is

$$21.16) \quad Median = L_1 + \left(\frac{\dfrac{N}{2} - \left(\sum f\right)_1}{f_{median}}\right) \times c$$

L_1 = lower class boundary of the class containing median (called median class)
N = total frequency, i.e., total number of observations
$(\sum f)_1$ = sum of the frequencies of all classes lower than the median class
f_{median} = frequency of the median class
c = size of the median class interval

21.3 Mode

The mode of N numbers such as $X_1, X_2, X_3, \ldots, X_N$ is

21.17) The value which occurs the most, i.e., having the greatest frequency. Mode can be more than one value.

Mode of grouped data (data arranged in a frequency distribution) is

21.18) $Mode = L_1 + \left(\dfrac{\Delta_1}{\Delta_1 + \Delta_2} \right) \times c$

L_1 = Lower class boundary of the class containing mode (called model class)
Δ_1 = excess of modal frequency over frequency of next lower class, i.e., difference between modal frequency and the frequency of the previous class
Δ_2 = excess of modal frequency over frequency of next higher class, i.e., difference between modal frequency and the frequency of the subsequent class
c = size of the modal class interval

21.4 Measures of Dispersion

21.4.1 Range

21.19) The difference between the largest and the smallest numbers in N numbers such as $X_1, X_2, X_3, \ldots, X_N$.

21.4.2 Mean or Average Deviation

For N numbers such as $X_1, X_2, X_3, \ldots, X_N$, mean deviation is

21.20) $MD = \dfrac{\sum\limits_{i=1}^{i=N} |X_i - \overline{X}|}{N}$ See Eq. 21.1

If numbers $X_1, X_2, X_3, \ldots, X_N$ occur $f_1, f_2, f_3, \ldots, f_N$ times, respectively, then their mean deviation is

21.21) $MD = \dfrac{\sum\limits_{i=1}^{i=N} f_i |X_i - \overline{X}|}{N}$ See Eq. 21.3

21.4.3 Standard Deviation

For N numbers such as $X_1, X_2, X_3, \ldots, X_N$, population standard deviation is

21.22) $\sigma = \sqrt{\dfrac{\sum\limits_{i=1}^{i=N}(X_i - \bar{X})^2}{N}}$ See Eq. 21.1

If numbers $X_1, X_2, X_3, \ldots, X_N$ occur $f_1, f_2, f_3, \ldots, f_N$ times respectively, then their population standard deviation is

21.23) $\sigma = \sqrt{\dfrac{\sum\limits_{i=1}^{i=N} f_i(X_i - \bar{X})^2}{N}}$ See Eq. 21.3

If a sample of data points is taken from the whole lot, then standard deviation of the whole lot is sample standard deviation, which is

21.24) $s = \sqrt{\dfrac{\sum\limits_{i=1}^{i=N}(X_i - \bar{X})^2}{N-1}}$ See Eq. 21.1

If numbers $X_1, X_2, X_3, \ldots, X_N$ occur $f_1, f_2, f_3, \ldots, f_N$ times respectively, then their sample standard deviation is

21.25) $s = \sqrt{\dfrac{\sum\limits_{i=1}^{i=N} f_i(X_i - \bar{X})^2}{N-1}}$ See Eq. 21.3

21.4.4 Variance

21.26) variance = (standard deviation)^2

For N numbers such as $X_1, X_2, X_3, \ldots, X_N$, population variance is

21.27) $\sigma^2 = \dfrac{\sum_{i=1}^{i=N}(X_i - \bar{X})^2}{N}$ See Eq. 21.1

If numbers $X_1, X_2, X_3, \ldots, X_N$ occur $f_1, f_2, f_3, \ldots, f_N$ times respectively, then their population variance is

21.28) $\sigma^2 = \dfrac{\sum_{i=1}^{i=N} f_i(X_i - \bar{X})^2}{N}$ See Eq. 21.3

For N numbers such as $X_1, X_2, X_3, \ldots, X_N$, sample variance is

21.29) $s^2 = \dfrac{\sum_{i=1}^{i=N}(X_i - \bar{X})^2}{N-1}$ See Eq. 21.1

If numbers $X_1, X_2, X_3, \ldots, X_N$ occur $f_1, f_2, f_3, \ldots, f_N$ times respectively, then their sample variance is

21.30) $s^2 = \dfrac{\sum_{i=1}^{i=N} f_i(X_i - \bar{X})^2}{N-1}$ See Eq. 21.3

21.4.5 Standard Error of Mean

21.31) Standard error of mean $= \dfrac{\text{standard deviation}}{\sqrt{\text{number of data points}}}$

21.32) $SE_M = \dfrac{\sigma}{\sqrt{N}}$ See Eq. 21.22

21.33) $SE_M = \dfrac{s}{\sqrt{N}}$ See Eq. 21.24

21.5 Measures of Error of Estimation

21.5.1 Sum of Squares of the Errors (*SSE*)

For N data pairs of experimental and model values:

21.34) $SSE = \sum_{i=1}^{i=N}(X_{obs,i} - X_{mod,i})^2$

where $X_{obs,i}$ is observed or experimental value and $X_{mod,i}$ is value calculated by a model.

21.5.2 R^2

For N data pairs of experimental and model values:

21.35) $R^2 = 1 - \dfrac{SSE}{SSM}$ See Eq. 21.34

21.36) $SSM = \sum_{i=1}^{i=N}(X_{obs,i} - \overline{X}_{obs})^2$ See Eq. 21.1

where $\overline{X}_{obs}$ is mean of observed or experimental values.

21.5.3 AdjR^2

For N data pairs of experimental and model values:

21.37) $\text{Adj}R^2 = 1 - \dfrac{SSE(N-1)}{SSM(N-m-1)} = 1 - \dfrac{(1-R^2)(N-1)}{N-m-1}$ See Eqs. 21.34 and 21.36

where m = number of parameters.

21.5.4 *F*-value

For N data pairs of experimental and model values:

21.38) $F = \dfrac{MSR}{MSE}$

21.39) $MSR = \dfrac{SSM - SSE}{m-1}$ See Eqs. 21.34 and 21.36

21.40) $MSE = \dfrac{SSE}{N-m}$

where m = number of parameters.

21.5.5 Root Mean Square Deviation

For N data pairs of experimental and model values:

21.41) $$RMSD = \sqrt{\frac{\sum_{i=1}^{i=N}(X_{obs,i} - X_{mod,i})^2}{N}} = \sqrt{\frac{SSE}{N}}$$

21.5.6 Standard Error of Estimate

For N data pairs of experimental and model values:

21.42) $$SE_{yx} = \sqrt{\frac{\sum_{i=1}^{i=N}(X_{obs,i} - X_{mod,i})^2}{N-2}} = \sqrt{\frac{SSE}{N-2}}$$

21.6 Curve Fitting Models

	Model	Equation
21.43	Linear	$y = ax + b$
21.44	Multiple linear (for 3 independent variables)	$y = a + bx_1 + cx_2 + dx_3$
21.45	Quadratic	$y = ax^2 + bx + c$
21.46	Cubic	$y = ax^3 + bx^2 + cx + d$
21.47	Higher degree polynomial	$y = a_0 x^n + a_1 x^{n-1} + a_2 x^{n-2} + \cdots + a_{n-1} x + a_n$
21.48	Reciprocal	$y = a + \dfrac{b}{x}$
21.49	Exponential growth	$y = ab^x \quad (\ln y = x \ln b + \ln a)$
21.50		$y = ae^{bx} \quad (\ln y = bx + \ln a)$
21.51	Exponential decay	$y = ae^{-bx} \quad (\ln y = -bx + \ln a)$
21.52	Exponential rise to the maximum	$y = a(1 - e^{-bx})$
21.53	Power	$y = ax^b \quad (\ln y = b \ln x + \ln a)$
21.54	Logarithmic	$y = a + b \log(x)$
21.55		$y = a + b \ln(x)$
21.56	Logistic (Sigmoid)	$y = \dfrac{1}{1 + e^{-x}}$

21.7 Statistical Distributions

	Name	Probability Density Function
21.57	Normal	$f(x) = \dfrac{1}{\sigma\sqrt{2\pi}} \exp\left(-\dfrac{(x-\mu)^2}{2\sigma^2}\right)$
21.58	Lognormal	$f(x) = \dfrac{1}{x\sigma_{\ln}\sqrt{2\pi}} \exp\left(-\dfrac{(\ln x - \mu_{\ln})^2}{2\sigma_{\ln}^2}\right)$
21.59	Gamma	$f(x) = \dfrac{1}{\beta^\alpha \Gamma(\alpha)} x^{\alpha-1} \exp\left(-\dfrac{x}{\beta}\right)$
21.60	Weibull	$f(x) = \dfrac{\alpha}{\beta}\left(\dfrac{x}{\beta}\right)^{\alpha-1} \exp\left(-\left(\dfrac{x}{\beta}\right)^\alpha\right)$

22.0 Mathematical Functions

22.1 Gamma, Beta, Error, and Complementary Error Functions

	Function	Form	Function	Form
22.1 22.2	Gamma	$\Gamma(x) = \int_0^\infty t^{x-1} \exp(-t) dt$	Error	$\mathrm{erf}(x) = \dfrac{2}{\sqrt{\pi}} \int_0^x \exp(-t^2) dt$
22.3 22.4	Beta	$B(x,y) = \int_0^1 t^{x-1}(1-t)^{y-1} dt$	Complementary error	$\mathrm{erfc}(x) = \dfrac{2}{\sqrt{\pi}} \int_x^\infty \exp(-t^2) dt$

22.5) $\mathrm{erfc}(x) = 1 - \mathrm{erf}(x)$

22.2 Values of Error Function

x	erf(x)	x	erf(x)	x	erf(x)	x	erf(x)
0	0.0000	0.38	0.4090	0.76	0.7175	1.18	0.9048
0.02	0.0226	0.40	0.4284	0.78	0.7300	1.22	0.9155
0.04	0.0451	0.42	0.4475	0.80	0.7421	1.26	0.9252
0.06	0.0676	0.44	0.4662	0.82	0.7538	1.30	0.9340
0.08	0.0901	0.46	0.4847	0.84	0.7651	1.40	0.9523
0.1	0.1125	0.48	0.5027	0.86	0.7761	1.50	0.9661
0.12	0.1348	0.50	0.5205	0.88	0.7867	1.60	0.9763
0.14	0.1569	0.52	0.5379	0.90	0.7969	1.70	0.9838
0.16	0.1790	0.54	0.5549	0.92	0.8068	1.80	0.9891
0.18	0.2009	0.56	0.5716	0.94	0.8163	1.90	0.9928
0.20	0.2227	0.58	0.5879	0.96	0.8254	2.0	0.9953
0.22	0.2443	0.60	0.6039	0.98	0.8342	2.1	0.9970
0.24	0.2657	0.62	0.6194	1.0	0.8427	2.2	0.9981
0.26	0.2869	0.64	0.6346	1.02	0.8508	2.3	0.9989
0.28	0.3079	0.66	0.6494	1.04	0.8586	2.4	0.9993
0.30	0.3286	0.68	0.6638	1.06	0.8661	2.5	0.9996
0.32	0.3491	0.70	0.6778	1.08	0.8733	2.6	0.9998
0.34	0.3694	0.72	0.6914	1.10	0.8802	2.7	0.9999
0.36	0.3893	0.74	0.7047	1.14	0.8931	2.8	0.9999

23.0 Numerical Mathematics

23.1 Interpolation
23.1.1 Linear

23.1) $f(x) = f(x_1) + \dfrac{f(x_2) - f(x_1)}{x_2 - x_1}(x - x_1)$

23.1.2 Quadratic

23.2) $f(x) = a_0 + a_1 x + a_2 x^2$

23.3) $a_0 = b_0 - b_1 x_0 + b_2 x_0 x_1$

23.4) $a_1 = b_1 - b_2 x_0 - b_2 x_1$

23.5) $a_2 = b_2$

23.6) $b_0 = f(x_0)$

23.7) $b_1 = \dfrac{f(x_1) - f(x_0)}{x_1 - x_0}$

23.8) $b_2 = \dfrac{\dfrac{f(x_2) - f(x_1)}{x_2 - x_1} - \dfrac{f(x_1) - f(x_0)}{x_1 - x_0}}{x_2 - x_0}$

23.1.3 Lagrange

23.9) $f_n(x) = \sum\limits_{i=0}^{n} L_i(x) f(x_i)$

23.10) $L_i(x) = \prod\limits_{\substack{j=0 \\ j \neq i}}^{n} \dfrac{x - x_j}{x_i - x_j}$

23.1.3.1 First Order Lagrange Polynomial

23.11) $f_1(x) = \dfrac{x - x_1}{x_0 - x_1} f(x_0) + \dfrac{x - x_0}{x_1 - x_0} f(x_1)$

23.1.3.2 Second Order Lagrange Polynomial

23.12) $f_2(x) = \dfrac{(x-x_1)(x-x_2)}{(x_0-x_1)(x_0-x_2)} f(x_0) + \dfrac{(x-x_0)(x-x_2)}{(x_1-x_0)(x_1-x_2)} f(x_1) + \dfrac{(x-x_0)(x-x_1)}{(x_2-x_0)(x_2-x_1)} f(x_2)$

23.2 Solution of Algebraic Equations

23.2.1 Newton-Raphson Method

For the equation $y = f(x)$, the root (solution) of the equation (x_n) can be approximated as

23.13) $x_{n+1} = x_n - \dfrac{f(x_n)}{f'(x_n)}$

23.2.2 Secant Method

For the equation $y = f(x)$, the root (solution) of the equation (x_n) can be approximated as

23.14) $x_{n+1} = x_n - \dfrac{x_n - x_{n-1}}{f(x_n) - f(x_{n-1})} f(x_n)$

23.3 Solution of First Order Ordinary Differential Equations (ODEs)

23.3.1 Euler Method

For the differential equation $\dfrac{dy}{dx} = f(x,y)$, the solution is

23.15) $y_{n+1} = y_n + h f(x_n, y_n)$
23.16) $h = x_{n+1} - x_n$

23.3.2 Improved Euler Method: Heun Method

For the differential equation $\dfrac{dy}{dx} = f(x,y)$, the solution is

23.17) $y^*_{n+1} = y_n + h f(x_n, y_n)$

23.18) $y_{n+1} = y_n + h \dfrac{f(x_n, y_n) + f(x_{n+1}, y^*_{n+1})}{2}$

23.16) $h = x_{n+1} - x_n$

23.3.3 Runge-Kutta (RK), 4th Order

For the differential equation $\frac{dy}{dx} = f(x, y)$, the solution is

23.19) $y_{n+1} = y_n + \frac{h}{6}(k_1 + 2k_2 + 2k_3 + k_4)$

23.16) $h = x_{n+1} - x_n$

23.20) $k_1 = f(x_n, y_n)$

23.21) $k_2 = f(x_n + \frac{1}{2}h, y_n + \frac{1}{2}k_1 h)$

23.22) $k_3 = f(x_n + \frac{1}{2}h, y_n + \frac{1}{2}k_2 h)$

23.23) $k_4 = f(x_n + h, y_n + k_3 h)$

23.3.4 Runge-Kutta-Fehlberg (RKF) Method

The Cash-Karp version of the Runge-Kutta-Fehlberg (RKF-45) method is given below (Ref.: Chapra, S.C.; Canale, R.P. 2001. Numerical Methods for Engineers. 4th ed. McGraw-Hill, New York, p. 719).

For the differential equation $\frac{dy}{dx} = f(x, y)$, the solution is

23.24) $y_{n+1} = y_n + h\left(\frac{37}{378}k_1 + \frac{250}{621}k_3 + \frac{125}{594}k_4 + \frac{512}{1771}k_6\right)$

23.25) $y_{n+1} = y_n + h\left(\frac{2825}{27648}k_1 + \frac{18575}{48384}k_3 + \frac{13525}{55296}k_4 + \frac{277}{14336}k_5 + \frac{1}{4}k_6\right)$

23.16) $h = x_{n+1} - x_n$

23.20) $k_1 = f(x_n, y_n)$

23.26) $k_2 = f\left(x_n + \frac{1}{5}h, y_n + \frac{1}{5}k_1 h\right)$

23.27) $k_3 = f\left(x_n + \dfrac{3}{10}h, y_n + \dfrac{3}{40}k_1h + \dfrac{9}{40}k_2h\right)$

23.28) $k_4 = f\left(x_n + \dfrac{3}{5}h, y_n + \dfrac{3}{10}k_1h - \dfrac{9}{10}k_2h + \dfrac{6}{5}k_3h\right)$

23.29) $k_5 = f\left(x_n + h, y_n - \dfrac{11}{54}k_1h + \dfrac{5}{2}k_2h - \dfrac{70}{27}k_3h + \dfrac{35}{27}k_4h\right)$

23.30) $k_6 = f\left(x_n + \dfrac{7}{8}h, y_n + \dfrac{1631}{55296}k_1h + \dfrac{175}{512}k_2h + \dfrac{575}{13824}k_3h + \dfrac{44275}{110592}k_4h + \dfrac{253}{4096}k_5h\right)$

23.4 Differentiating a Function: Finite Differences

(Ref.: Chapra, S.C.; Canale, R.P. 2001. Numerical Methods for Engineers. 4th ed. McGraw-Hill, New York, p. 633)

23.4.1 Forward Finite Differences

	Lower Order Accuracy
First derivative	$f'(x) = \dfrac{f(x_{n+1}) - f(x_n)}{h}$
Second derivative	$f''(x) = \dfrac{f(x_{n+2}) - 2f(x_{n+1}) + f(x_n)}{h^2}$
	Higher Order Accuracy
First derivative	$f'(x) = \dfrac{-f(x_{n+2}) + 4f(x_{n+1}) - 3f(x_n)}{2h}$
Second derivative	$f''(x) = \dfrac{-f(x_{n+3}) + 4f(x_{n+2}) - 5f(x_{n+1}) + 2f(x_n)}{h^2}$

23.4.2 Backward Finite Differences

	Lower Order Accuracy
First derivative	$f'(x) = \dfrac{f(x_n) - f(x_{n-1})}{h}$
Second derivative	$f''(x) = \dfrac{f(x_n) - 2f(x_{n-1}) + f(x_{n-2})}{h^2}$
	Higher Order Accuracy
First derivative	$f'(x) = \dfrac{3f(x_n) - 4f(x_{n-1}) + f(x_{n-2})}{2h}$
Second derivative	$f''(x) = \dfrac{2f(x_n) - 5f(x_{n-1}) + 4f(x_{n-2}) - f(x_{n-3})}{h^2}$

23.4.3 Centered Finite Differences

	Lower Order Accuracy
First derivative	$f'(x) = \dfrac{f(x_{n+1}) - f(x_{n-1})}{2h}$
Second derivative	$f''(x) = \dfrac{f(x_{n+1}) - 2f(x_n) + f(x_{n-1})}{h^2}$
	Higher Order Accuracy
First derivative	$f'(x) = \dfrac{-f(x_{n+2}) + 8f(x_{n+1}) - 8f(x_{n-1}) + f(x_{n-2})}{12h}$
Second derivative	$f''(x) = \dfrac{-f(x_{n+2}) + 16f(x_{n+1}) - 30f(x_n) + 16f(x_{n-1}) - f(x_{n-2})}{12h^2}$

23.5 Integrating a Function

23.5.1 Trapezoidal Rule

23.31) $\int_a^b f(x)dx = \dfrac{h}{2}\left(f(x_0) + 2f(x_1) + 2f(x_2) + \cdots + 2f(x_{n-1}) + f(x_n)\right)$

23.32) $h = \dfrac{b-a}{n}$

Equal sized intervals are required to be used.

23.5.2 Simpson 1/3 Rule

23.33) $\int_a^b f(x)dx = \dfrac{h}{3}\left(f(x_0) + 4f(x_1) + 2f(x_2) + 4f(x_3) + \cdots + 2f(x_{n-2}) + 4f(x_{n-1}) + f(x_n)\right)$

23.32) $h = \dfrac{b-a}{n}$

There should be even number of intervals of equal size.

23.6 Partial Differential Equations

23.6.1 Classification of Second Order Linear Partial Differential Equations

A linear second order partial differential equation with three independent variables can be generally written as

23.34) $A\dfrac{\partial^2 T}{\partial x^2} + B\dfrac{\partial^2 T}{\partial x \partial y} + C\dfrac{\partial^2 T}{\partial y^2} + D\dfrac{\partial T}{\partial x} + E\dfrac{\partial T}{\partial y} + FT + G = 0$

The above equation (Eq. 23.34) can be classified, based on the value of $B^2 - 4AC$, into three types of commonly appeared differential equations:

	$B^2 - 4AC$	Type	Example
23.35	< 0	Elliptical	Laplace equation: $\dfrac{\partial^2 T}{\partial x^2} + \dfrac{\partial^2 T}{\partial y^2} = 0$
23.36	$= 0$	Parabolic	Diffusion equation: $\dfrac{\partial T}{\partial t} = \alpha \left(\dfrac{\partial^2 T}{\partial x^2} + \dfrac{\partial^2 T}{\partial y^2} \right)$
23.37	> 0	Hyperbolic	Wave equation: $\dfrac{\partial^2 T}{\partial x^2} + \dfrac{\partial^2 T}{\partial y^2} = \dfrac{1}{c^2} \dfrac{\partial^2 T}{\partial t^2}$

23.6.2 Explicit Finite Difference Method for One-Dimensional Parabolic Differential Equation

A one-dimensional (spatial dimension) parabolic differential equation (diffusion equation) is given below.

23.38) $\dfrac{\partial T}{\partial t} = \alpha \dfrac{\partial^2 T}{\partial x^2}$

The spatial or second derivative approximation, by centered finite difference approach, for Eq. 23.38 is

23.39) $\dfrac{\partial^2 T}{\partial x^2} = \dfrac{T^n_{m+1} - 2T^n_m + T^n_{m-1}}{(\Delta x)^2}$ See Section 23.4

The time or first derivative approximation, by forward finite difference approach, for Eq. 23.38 is

23.40) $\dfrac{\partial T}{\partial t} = \dfrac{T_m^{n+1} - T_m^n}{\Delta t}$ See Section 23.4

Eqs. 23.38, 23.39, and 23.40 yield

23.41) $\dfrac{T_m^{n+1} - T_m^n}{\Delta t} = \alpha \dfrac{T_{m+1}^n - 2T_m^n + T_{m-1}^n}{(\Delta x)^2}$

23.42) $T_m^{n+1} = T_m^n + \dfrac{\alpha \Delta t}{(\Delta x)^2}(T_{m+1}^n - 2T_m^n + T_{m-1}^n)$

A small value of $\dfrac{\alpha \Delta t}{(\Delta x)^2} \leq 0.25$ may well result in a stable non-oscillatory solution. Eq. 23.42 is applicable to all interior points of the finite difference grid. At the edges, boundary conditions have to be specified. If T is specified, say, $T(x)$ is provided, it is Dirichlet boundary condition. If derivative of T normal to the boundary is specified, say, $\dfrac{\partial T}{\partial x}$ is provided, it is called Neumann boundary condition. A combination of both of these boundary conditions can also be specified.

24.0 Selected MATLAB Commands with Examples

(MATLAB® is registered trademark of The MathWorks, Inc.)

	Purpose	Command
24.1	π	`>>pi`
24.2	Infinity	`>>inf`
24.3	2.371×10^5	`>>2.371e5`
24.4	Format for 15 digits after decimal in the output	`>>format long`
24.5	Format for 4 digits after decimal in the output	`>>format short`
24.6	Basic operations: $2 \times 7 + 3 - 5 + 3^\wedge 2 + 15/3$	`>>2*7+3-5+3^2+15/3`
24.7	Absolute value of a number	`>>abs(-3)`
24.8	Square root of a number	`>>sqrt(13)`
24.9	Exponential (e^x) of a number	`>>exp(3)`
24.10	Natural logarithm of a number	`>>log(10)`
24.11	Logarithm to the base 2 of a number	`>>log2(10)`
24.12	Logarithm to the base 10 of a number	`>>log10(10)`
24.13	Factorize a number	`>>factor(256)`
24.14	Factorial of a number	`>>factorial(5)`
24.15	Remainder after division of two numbers	`>>rem(5,4)`
24.16	sin, cos, and tan of angles in radian	`>>sin(pi)`
24.17	sin, cos, and tan of angles in degree	`>>sind(180)`
24.18	Inverse sin, cos, and tan of angles in radian	`>>asin(1)`
24.19	Inverse sin, cos, and tan of angles in degree	`>>asind(90)`
24.20	sin, cos, and tan hyperbolic	`>>sinh(pi)`
24.21	Inverse sin, cos, and tan hyperbolic	`>>asinh(1)`
24.22	Vector of 1 to 100 with interval of 1	`>>P=[1:100];`
24.23	Vector of 1 to 100 with interval of 10	`>>Q=[1:10:100];`
24.24	Vector of 100 data points from 1 to 10	`>>R=linspace(1,10)`
24.25	Vector of 10 data points from 1 to 100	`>>S=linspace(1,100,10)`
24.26	Assigning given data to variable T (in a row vector)	`>>T=[10,20,30,40,50,60,70];` or `>>T=[10 20 30 40 50 60 70];`
24.27	Assigning given data to variable T1 (in a column vector)	`>>T1=[10;20;30;40;50;60;70];` or `>>T1=T'` (See Eq. 24.35) or `>>T1=[10` `20` `30` `40 press "enter" after each and close with ];`

24.28	Finding number of elements of a vector	>>length(T)	See Eq. 24.26
24.29	Sum of elements of a vector	>>sum(T)	See Eq. 24.26
24.30	Product of elements of a vector	>>prod(T)	See Eq. 24.26
24.31	Maximum value in a vector	>>max(T)	See Eq. 24.26
24.32	Minimum value in a vector	>>min(T)	See Eq. 24.26
24.33	Mean or average (arithmetic) of data	>>mean(T)	See Eq. 24.26
24.34	Sorting of elements of a vector	>>U=[2,3,5,1,9,7,4]; >>sort(U)	
24.35	Transpose of a row vector (row vector to column vector)	>>U1=U'	See Eq. 24.34
24.36	Specific value in a vector (3rd value)	>>T(3)	See Eq. 24.26
24.37	Add scalar to elements of a vector	>>T+273.15	See Eq. 24.26
24.38	Multiply scalar with elements of a vector	>>1.5*T	See Eq. 24.26
24.39	Difference between adjacent elements of a vector	>>V=[2,3,3,4,7,9,13]; >>diff(V)	
24.40	Add two vectors and store the resultant values to W	>>W=T+V See Eqs. 24.26 and 24.39	
24.41	Subtract two vectors	>>T-V	See Eqs. 24.26 and 24.39
24.42	Multiply two vectors	>>T.*V	See Eqs. 24.26 and 24.39
24.43	Divide two vectors (element by element)	>>T./V	See Eqs. 24.26 and 24.39
24.44	Mode of data	>>mode(V)	See Eq. 24.39
24.45	Median of data	>>median(V)	See Eq. 24.39
24.46	Variance of data	>>var(V)	See Eq. 24.39
24.47	Standard deviation of data	>>std(V)	See Eq. 24.39
24.48	Dot and cross product of two vectors $\vec{A} = 2\hat{i} + 3\hat{j} + 5\hat{k}$ and $\vec{B} = 3\hat{i} + 7\hat{j} + 11\hat{k}$	>>A=[2 3 5]; B=[3 7 11]; >>dot(A,B) >>cross(A,B)	
24.49	Plotting (2-D, "o" symbol, and solid line) Naming of x- and y-axes Showing legend Drawing grid	>>plot(T,V,'o-') >>xlabel('temp, C'),ylabel('vol, m3') >>legend('T vs V') >>grid on See Eqs. 24.26 and 24.39	
24.50	log-log (2-D) plotting	>>loglog(T,V) See Eqs. 24.26 and 24.39	
24.51	Semilog (2-D) plotting with log values along x-axis Semilog (2-D) plotting with log values along y-axis	>>semilogx(T,V) >>semilogy(T,V) See Eqs. 24.26 and 24.39	
24.52	Plotting (3-D, "o" symbol, solid line, and grid) Naming of x-, y-, and z-axes	>>plot3(T,V,W,'o-'),grid on >>xlabel('T'),ylabel('V'),zlabel('W') See Eqs. 24.26, 24.39, and 24.40	

24.53	Linear fitting of data
	2nd degree polynomial fitting of data
	3rd degree polynomial fitting of data
	Fitting of data with function $a\exp(bx)$
	Fitting of data with function ax^b
	Fitting of data with function ax^b+c
		`>>polyfit(T,U,1)`
		`>>polyfit(T,U,2)`
		`>>polyfit(T,U,3)`
		`>>fit(T1,U1,'exp1')`
		`>>fit(T1,U1,'power1')`
		`>>fit(T1,U1,'power2')`
		See Eqs. 24.26, 24.27, 24.34, and 24.35

24.54	Matrix of 3×3 order	`>>M=[2,3,3; 9,5,6; 3,7,11];`
24.55	Identity matrix of 3×3 order	`>>I=eye(3)`
24.56	Null matrix of 3×3 order	`>>zeros(3,3)`
24.57	Matrix of random numbers of 3×3 order	`>>N=rand(3,3)`
24.58	Specific value in a matrix (8th member)	`>>M(8)` or `>>M(2,3)` See Eq. 24.54
24.59	Finding size of a matrix	`>>size(M)` See Eq. 24.54
24.60	Determinant of a matrix	`>>det(M)` See Eq. 24.54
24.61	Rank of a matrix	`>>rank(M)` See Eq. 24.54
24.62	Multiplication of a matrix by a scalar	`>>P=1.5*M` See Eq. 24.54
24.63	Sum of two matrices of 3×3 order	`>>P+M` See Eqs. 24.54 and 24.62
24.64	Subtraction of two matrices of 3×3 order	`>>P-M` See Eqs. 24.54 and 24.62
24.65	Multiplication of two matrices of 3×3 order	`>>M*P` See Eqs. 24.54 and 24.62
24.66	Multiplication of two matrices of 3×3 order (element by element)	`>>M.*P` See Eqs. 24.54 and 24.62
24.67	Transpose of a matrix	`>>M1=M'` See Eq. 24.54
24.68	Inverse of a matrix	`>>inv(M)` See Eq. 24.54

24.69	Solving simultaneous linear algebraic equations using matrix method
	$2x+3y+5z=6$
	$3x+y+9z=7$
	$5x+7y+6z=13$
		`>>A=[2, 3, 5; 3, 1, 9; 5, 7, 6];`
		`>>B=[6; 7; 13];`
		`>>X=inv(A)*B` or
		`>>X=A\B` or
		`>>X=B'/A'` or
		`>>X=linsolve(A,B)`

24.70	Finding roots of the polynomial $x^2+3x+2=0$	`>>p=[1, 3, 2];` `>>roots(p)`
24.71	Finding coefficient of the polynomial with roots −2 and −1	`>>r=[-2,-1]` `>>poly(r)`
24.72	Evaluation of the polynomial $x^2+3x+2=0$ at given value of $x=2$	`>>polyval(p,2)` See Eq. 24.70
24.73	Derivative of the polynomial $x^2+3x+2=0$	`>>polyder(p)` See Eq. 24.70
24.74	Function with name f1 (M-file with name f1)	`function y=f1(x)` `y=x^2+3*x+2;`

24.75	Inline function f2. f2 function is stored so that it can be called any time. No need to write an M-file.	`>>f2=inline('x^2 +3*x +2','x')` or `>>f2=@(x) x^2+3*x+2`
24.76	Solving single algebraic equation $x^2+3x+2=0$ with initial guess of 2.0	`>>x=fzero(@f1, 2)` See Eq. 24.74
24.77	Finding function value at a given point	`>>y=feval(@f1,3)` or `>>y=feval('f1',3)` See Eq. 24.74
24.78	Solving simultaneous linear or non-linear algebraic equations $2x+3y+5z=6$ $3x+y+9z=7$ $5x+7y+6z=13$	`function n=f3(x)` `n=[2*x(1)+3*x(2)+5*x(3)-6;` `3*x(1)+x(2)+9*x(3)-7;` `5*x(1)+7*x(2)+6*x(3)-13];` `>>x0=[1, 1, 1];` (initial guess) `>>fsolve('f3',x0)`
24.79	Sum of the series $\sum_{i=1}^{N=1}\frac{1}{x^2}$ for the first 10 terms	`>>syms x` `>>s=symsum((1/x^2),1,10)` `>>eval(s)`
24.80	Taylor series of $\log(1-x)$ for the first 6 terms about $a=0$	`>>syms x` `>>y=log(1-x);` `>>taylor(y,6,0)`
24.81	Numerical integration of $\int_1^{10}\frac{x}{2}dx$	`>>quad('x/2',1,10)` or `>>x=[1,2,3,4,5,6,7,8,9,10];` `>>y=x/2;` `>>trapz(x,y)`
24.82	Double integration of $\int_0^\pi\int_0^\pi(\sin x+\cos y)dxdy$	`>>f4=inline('sin(x)+cos(y)','x','y')` `>>dblquad(f4,0,pi,0,pi)`
24.83	Numerically solving the first order ODE $\frac{dy}{dx}-2x-4y=0$, $y(0)=1$, $x_1=0$, $x_2=0.5$	`function z=f5(x, y)` `z=2*x+4*y;` `>>[x y]=ode45('f5',[0,0.5],1)`
24.84	Simultaneous solution of the first order ODEs $\frac{dy_1}{dz}=y_2$, $y_1(0)=1$ $\frac{dy_2}{dz}=y_1-3$, $y_2(0)=0$	`>>f6=@(z,y) [y(2)-3;y(1)]` `>>[z,y]=ode45(f6,[0,0.25],[1;0])`
24.85	Factorize the function (a^3-b^3)	`>>syms a b` `>>factor(a^3-b^3)`
24.86	Expand the function $(a-b)(a^2+ab+b^2)$	`>>syms a b` `>>expand((a-b)*(a^2+a*b+b^2))`
24.87	Simplify the function $(a-b)(a^2+ab+b^2)$	`>>syms a b` `>>simplify((a-b)*(a^2+a*b+b^2))`
24.88	Symbolic solution of the algebraic equation $x^2+3x+2=0$	`>>solve('x^2+3*x+2')` or `>>syms x` `>>solve(x^2+3*x+2)`

24.89	Coefficients of a polynomial to its symbolic expression	>>poly2sym(p)　　See Eq. 24.70
24.90	Symbolic differentiation of x^2 with respect to x	>>diff('x^2','x') or >>syms x >>diff(x^2,'x')
24.91	Symbolic integration of $\int \frac{x}{2}dx$ Symbolic integration of $\int_1^{10} \frac{x}{2}dx$	>>int('x/2','x') or >>syms x >>int(x/2) >>int('x/2','x',1,10)
24.92	Symbolic solution of the differential equation $\frac{d^2y}{dx^2} - 2\frac{dy}{dx} - 4y = 0$ $\frac{d^2y}{dx^2} - 2\frac{dy}{dx} - 4y = 0$, $y(0) = 1$, $y'(0) = 2$	>>dsolve('D2y-2*Dy-4*y=0', 'x') >>dsolve('D2y-2*Dy-4*y=0','y(0)=1','Dy(0)=2', 'x')
24.93	Error function of a number	>>erf(0.02)
24.94	Complementary error function of a number	>>erfc(0.02)
24.95	Laplace transform of e^{at}	>>syms a t >>laplace(exp(a*t))
24.96	Inverse Laplace of $\frac{1}{s+a}$	>>syms a s >>ilaplace(1/(s+a))
24.97	Transfer function $\frac{5}{2\tau + 1}$	>>num=[5]; >>den=[2,1]; >>G=tf(num,den)
24.98	Step input	>>step(G)　　See Eq. 24.97
24.99	Impulse input	>>impulse(G)　　See Eq. 24.97
24.100	Partial fraction of the function $\frac{x+4}{x^2+3x+2}$	>>syms x >>H=(x+4)/(x^2+3*x+2); >>diff(int(H))　or >>a=[1 4]; b=[1 3 2]; >>[r, p, k]=residue(a,b)
24.101	Help on a command	>>doc plot >>help plot
24.102	Exit MATLAB	>>exit or >>quit

25.0 Suggested Sources

- https://www.wolframalpha.com [Last accessed on: 05-Sep-2017]
- http://mathworld.wolfram.com [Last accessed on: 05-Sep-2017]
- Gieck, K.; Gieck, R. 2006. Engineering Formulas. 8th American ed. McGraw-Hill Co. Inc.
- Spiegel, M.R.; Lipschutz, S.; Liu, J. 2009. Schaum's Outlines Mathematical Handbook of Formulas and Tables. 3rd ed. McGraw-Hill, New York.
- Chapra, S.C.; Canale, R.P. 2001. Numerical Methods for Engineers. 4th ed. McGraw-Hill, New York.

www.ingramcontent.com/pod-product-compliance
Lightning Source LLC
Chambersburg PA
CBHW082342220526
45470CB00008B/2604